GEORGIAN BAY

Discovering A Unique North American Ecosystem

Edited by Nick Eyles

Fitzhenry & Whiteside

Published in Canada by Fitzhenry & Whiteside Limited, 209 Wicksteed Avenue, Unit 51, Toronto, ON M4G 0B1.
Published in the United States by Fitzhenry & Whiteside Limited. 60 Leo M Birmingham Pkwy, Ste 107, Brighton, MA 02135

Fitzhenry & Whiteside acknowledges with thanks the Canada Council for the Arts and the Ontario Arts Council for their support of our publishing program. We acknowledge the financial support of the Government of Canada through the Canada Book Fund (CBF) for our publishing activities.

ONTARIO ARTS COUNCIL
CONSEIL DES ARTS DE L'ONTARIO
an Ontario government agency
un organisme du gouvernement de l'Ontario

Canada Council for the Arts | Conseil des arts du Canada

Library and Archives Canada Cataloguing in Publication

Title: Georgian Bay : discovering a unique North American ecosystem / edited by Nick Eyles.
Names: Eyles, Nick, 1952- editor.
Description: Paperback edition. | Includes bibliographical references and index.
Identifiers: Canadiana 20230154697 | ISBN 9781554556236 (softcover)
Subjects: LCSH: Natural history—Ontario—Georgian Bay (Bay) | LCSH: Biotic communities—Ontario—Georgian Bay (Bay) | LCSH: Ecosystem management—Ontario—Georgian Bay (Bay) | LCSH: Georgian Bay (Ont. : Bay)—Environmental conditions.
Classification: LCC QH77.C3 G47 2023 | DDC 577.09713/15—dc23

Publisher Cataloging-in-Publication Data (U.S.)

Names: Eyles, Nick, 1952- author.
Title: Georgian Bay : A Unique North American Ecosystem / Nick Eyles.
Description: Toronto, Ontario : Fitzhenry & Whiteside, 2023. | Includes bibliography and index. | Summary: "A project of the Georgian Bay Land Trust, this book focuses on the preservation of the unique features of the Georgian Bay ecosystem" – Provided by publisher.
Identifiers: ISBN 978-1-55455-381-5 (hardcover-2017) | 978-1-55455-623-6 (paperback)
Subjects: LCSH: Ecology – Canada – Georgian Bay (Ont.) | Ecosystem management – Canada – Georgian Bay (Ont.) | BISAC: SCIENCE / Life Sciences / Ecology.
Classification: LCC QH77.C2E954 |DDC 333.72 – dc23

Text and design by Kerry Designs
Cover image courtesy Kerry Plumley
Cover design by Tanya Montini
Printed in Canada by Marquis Book

www.fitzhenry.ca

FOREWORD

In late 2011 Nick Eyles penned and shared with a small group of his fellow Georgian Bay Land Trust Directors an outline of a book about Georgian Bay. The book would cover a broad range of topics about the Bay and we would need to find experts who might be able to help us with the various subjects which had been laid out. Now, six years later, this splendid book has been published.

The GBLT thanks all those involved in making this a reality: Nick Eyles for both his wonderfully written chapters along with the original concept; Sharon Fitzhenry for agreeing from the beginning to publish the book, Karl Schiefer, Susan Hannon, Bob Whittam, Peter Storck, Ron Williamson, Martin Cooper, Jamie Hunter, Andrew Miall, Christine Boyanoski and Bill Lougheed for their hard work and generous agreement to each write chapters in the book. Each one took time from their own work to help us out. The end result, we believe is a great contribution to those interested in this magnificent place: Georgian Bay.

This year, 2017, marks the first 26 years of existence for the Georgian Bay Land Trust. It has taken many people to accomplish what we have done to date. From those who had the foresight to start the Land Trust, to those who have donated lands, put easements on their properties, donated time and money, stewarded our properties, carried the message and continue to do so-we thank you. As it has often been said-it takes a village to do great things.

The Georgian Bay Land Trust dedicates this book to all those before us, with us now and those that will come after us. Those that have or will someday paddle the pristine cold waters in the early spring and watch the great blue herons rise up in flight from amongst the cattails in the marsh. That have sat or will sit on the lichen covered rocks to see the sun dip slowly on a warm summer night. That have or will love this Georgian Bay. And we hope that as we have learned from those before us to take care, to conserve, value and appreciate, so too can we teach our children and they theirs to do the same.

Janny Vincent,
Chair, Georgian Bay Land Trust

CONTENTS

On the rocks: a summer excursion hosted by the Georgian Bay Land Trust.

EDITOR'S NOTE

The Georgian Bay Land Trust is a not-for-profit charity registered in Ontario. Its mission is to preserve sites within the unique archipelago and the adjacent water bodies that lie along the eastern shore of Georgian Bay and the North Channel of Lake Huron that are of special ecological, geological and cultural importance. In 1976 Dr. Ross Lord, Dean of the University of Toronto's School of Practical Science and Founder of the Conservation Authority of Toronto, brought forward the notion of protecting that region's environment through a land trust–an organization that acts as steward to protect and preserve private land for the common public good. Incorporation and registration of the Georgian Bay Land Trust (GBLT) was completed in 1991. GBLT holds, stewards and protects in perpetuity some 2,000 acres of GBLT properties, partner properties, and other natural areas on both private and public land within the remarkable '30,000 Islands' area. This is the longest and most extensive freshwater archipelago anywhere in the world with a total coastal length of over 5,300 km (mainland and islands) and contains the most diverse coastal habitats anywhere in the Great Lakes.

The motivation behind this book project was the oft-repeated comment regarding the need to tell the story of how the special place known as Georgian Bay came to be. To create a book that does it full justice is somewhat daunting and we really hope this might be it. This volume simply wouldn't have happened without the thoughtful contributions and very considerable patience of our authors Karl Schiefer, Peter Storck, Ron Williamson, Martin Cooper, Christine Boyanoski, Susan Hannon, Bob Whittam, Andrew Miall, Jamie Hunter and Bill Lougheed. It has been a great pleasure working with them all. All royalties are being re-directed to further the amazing conservation work of the Georgian Bay Land Trust.

The ongoing encouragement of our publisher Sharon Fitzhenry and the personal financial support of Janny Vincent, Chair of the Georgian Bay Land Trust were absolutely crucial to keeping the project alive and moving along. The ever creative Kerry Plumley has made the book a visual treat and Richard Dionne furnished sterling editorial assistance and sage advice. We are very grateful to the following for suggestions, encouragement, sharing their adventures on the Bay, help in tracking down figures, drafting images and maps, and obtaining the necessary permissions; Natasha Tsakiris of Fitzhenry and Whiteside, Alan and Charlotte Stein, Michele Pacheco of Sears Canada, April Hawkins, Brian Boyle and Nicola Woods of the Royal Ontario Museum, Amy Sultana of the West Parry Sound District Museum, Anik Glaude of the Varley Art Gallery of Markham, Arthur Roberts of Simon Fraser University, Kate Russell of the South Grey Museum, Susan Warner and Melissa Shaw of the Collingwood Museum, Tracy Mallon-Jenson of the Art Gallery of Ontario, Marg Rutka of the Ontario Geological Survey, Julia Abraham of the University of Toronto at Scarborough, Raven Amiro of the National Gallery of Canada, Alexndra Cousins of the McMichael Collection of Canadian Art, Ursula Bondy, Alexandra Cousins, Cameron McCuaig, John Krasinski, Rick and Michele Doktor, Michael Allder, Ed Bartram, Nick Culshaw, Peter and Cathy Cooper, Janet Lougheed, Larry and Diana Thompson, Mike Doughty, Katherine Wallace, Peter Wood, Amalia Kolovos, Shane Sookhan, Lina Arbelaez Moreno, Kirsten Kennedy, and Carolyn, Claire and Chris Eyles.

Thanks everyone!
Nick Eyles
University of Toronto, January 2017

***Georgian Bay's west coast:** Halfway Log Dump Bay, Bruce Peninsula where the Niagara Escarpment drops into Georgian Bay.*

INTRODUCTION

NICK EYLES

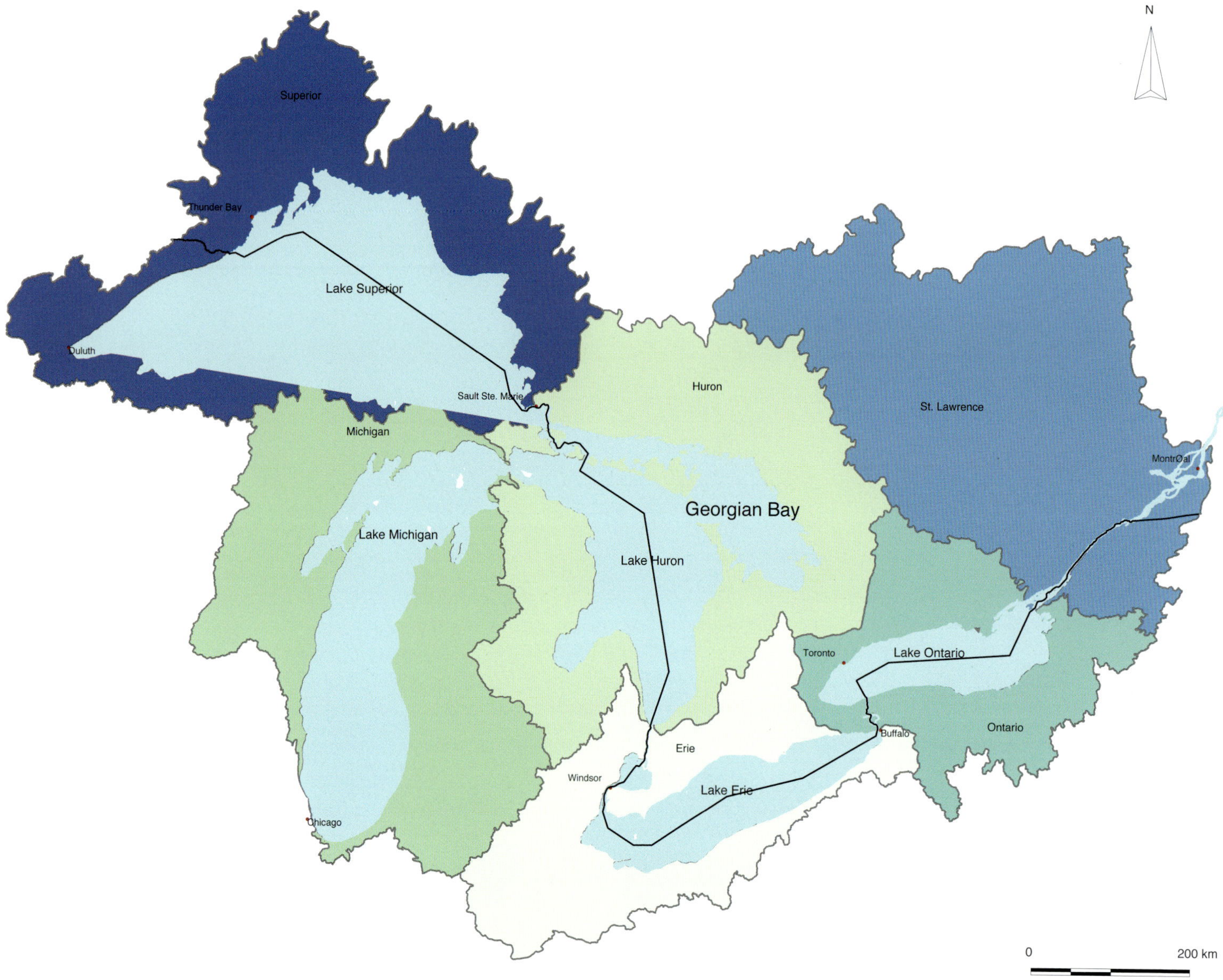

The five Great Lakes and their drainage basins *Georgian Bay is essentially the sixth Great Lake. It was named after King George IV of Great Britain.*

Right: ***Part of the 30,000 Islands area*** *of eastern Georgian Bay where the bumpy surface of the Canadian Shield becomes submerged under the water of Lake Huron.*

Georgian Bay (15,000 km^2) is the sixth Great Lake, often forgotten as such and the only one not named after indigenous peoples. It is a place that straddles major geologic, climatic, ecological and cultural boundaries; what has been called the 'land in between' along the southernmost margin of the seemingly sterile Precambrian rocks of the Canadian Shield with the much softer sedimentary rocks and farmlands to the south. As such Georgian Bay is a unique aquatic and terrestrial ecosystem.

Georgian Bay's west coast is dominated by the cliffed limestones of the Niagara Escarpment defining the backbone of the Bruce Peninsula and Manitoulin Island with prominent glacially-carved gaps allowing its waters to interchange with Lake Huron. The limestone plains of the Bay's west coast contrast with the 'Thirty Thousand Islands district' (there are closer to 100,000 islands at the last count) on its opposing eastern shore that form the most extensive freshwater archipelago anywhere in the world. Georgian Bay straddles the boundary between

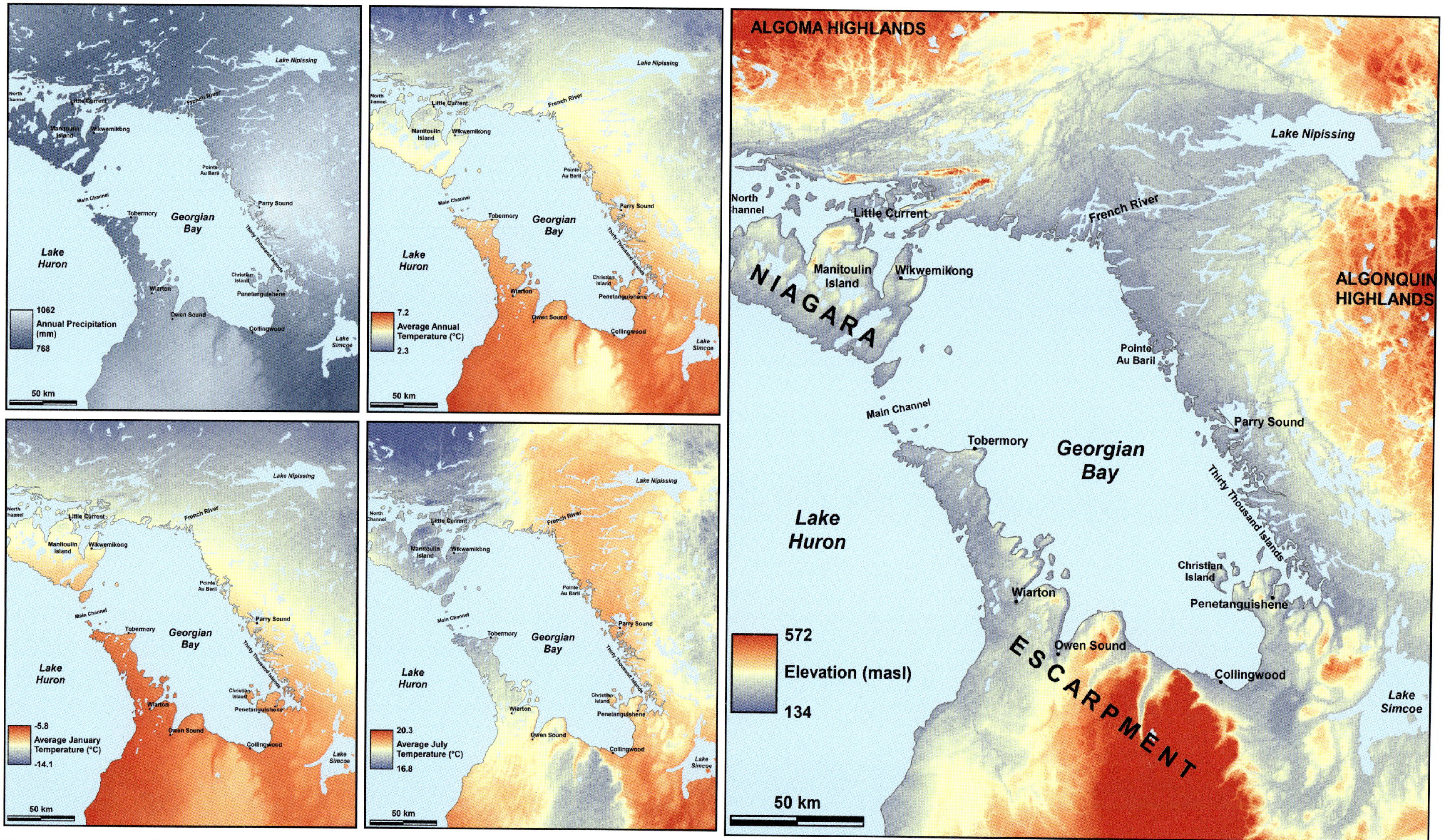

The topography of the Georgian Bay basin reflects its underlying geology with the highest areas along the Niagara Escarpment and across the Algoma and Algonquin highlands. The French River follows a broad topographic low forming a gateway through which ice from Quebec invaded Ontario during successive ice ages. This topography also controls modern day climate.

the boreal Shield to the north and the St.Lawrence Lowlands. It is what ecologists call an 'ecotone' sharing many elements of north and south, and is consequently more diverse than either.

The myriad islands and protected shallow inlets along the eastern shoreline of Georgian Bay, where the waters of Georgian Bay have flooded the bumpy, glacially-scoured surface of the Canadian Shield, contrasts greatly with the opposing coastline along the Bruce Peninsula dominated by the imposing Niagara Escarpment that drops precipitously into deeper open water.

This book

To date, there has been no systematic effort to describe the ecology, geology and landscapes of Georgian Bay and its aquatic and terrestrial ecosystems, its organisms and the history of human occupancy. Elizabeth Campbell lamented in '*The West Wind*' (published in 2005) on Georgian Bay that '*it seems odd that no one has asked why and how*

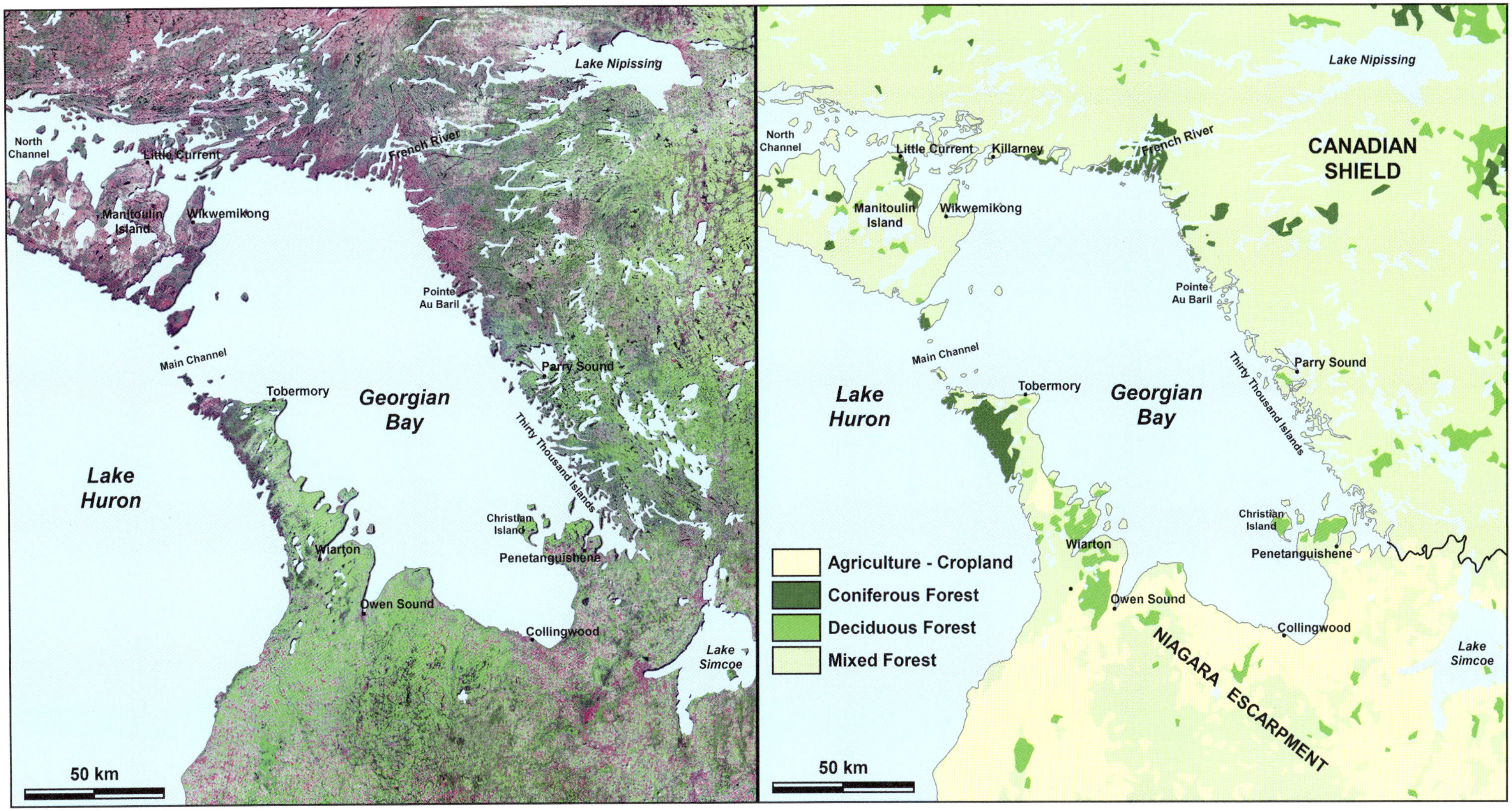

an iconic landscape was created.' The 25th Anniversary of the founding of the Georgian Bay Land Trust in 2016 has been the impetus for this book which brings together geologists, ecologists, historians, archeologists and artists to tell the unique story of Georgian Bay.

Starting 2 billion years ago and ending with the ice ages this book traces the origins of its rocks, landscapes, waters and ecosystems. Humans arrived from Asia 11,500 years ago just as the last ice sheet was receding northwards exposing more and more of the Bay. These skillful hunters formed a stable sustainable culture on the southern margins of the Canadian Shield that lasted some 9,000 years until their agricultural descendants built settled communities trading corn with northern hunter gatherers; the foundation of the fur trade after 1600. We emphasize in the pages that follow the profound impacts of accelerated European settlement after 1840, the challenges facing the early settlers, the impact of the railways and loggers and the great changes currently underway as the area moves away from being viewed as a resource to be exploited to a unique geological, ecological and cultural icon to be conserved. The book ends appropriately enough with a review of conservation measures taken over the last 100 years and outlines what needs to be done to keep this land an icon while balancing public accessibility with protection.

This then is the story of Georgian Bay.

North and South The varying colours of the map on the left portray the 'reflectivity' of the ground surface around Georgian Bay measured from satellites; green signifies cropland and forest cover whereas purple identifies much more reflective areas of exposed bare rock typical of the north on Manitoulin Island and the Shield. The same contrast is seen in land use (at right) which identifies cropland to the south from extensive forest cover of the north and east.

The Fountain of the Great Lakes by Lorado Taft erected in 1913 outside the Art Institute of Chicago. Water moves through the fountain in the same sequence as it moves through the Great Lakes; from an allegorical Lake Superior thence to Lakes Michigan, Huron, Erie, and Ontario who has her arms outstretched to the Atlantic Ocean. The statue was inspired by the widely acknowledged failure of the 1893 Columbian Exposition in Chicago to create a single sculpture emphasising the natural resources of the Great Lakes. Today, management of the Great Lakes and their unrivalled resource of freshwater is under the control of the International Joint Commission between Canada and the US. The waters of Georgian Bay, the sixth Great Lake, contains a globally unique freshwater archipelago: the Thirty Thousand Islands.

***The complex rocks of the Canadian Shield** that are well exposed along the eastern shores of Georgian Bay attract geologists from around the world as they comprise the billion years-old roots of former mountains. Banded gneisses seen here have been warmed, stretched and folded ('metamorphosed') as much as 25 km deep below the Earth's surface. These complexly deformed rocks allow geologists to reconstruct processes at work in deep Earth environments otherwise unobservable and fascinate artists with their Pollock-like appearance.*

CHAPTER

1

GEORGIAN BAY ROCKS: TWO BILLION YEARS IN THE MAKING

NICK EYLES

The white quartzites of the Lorrain Formation *form the backbone of the La Cloche Range seen here north of Manitoulin Island. These are the oldest rocks in the Georgian Bay region and are about 2.4 billion years old and were folded and bent some 1.8 billion years ago. These were quarried by Paleo-Indians about 11,000 years ago for stone tools.*

To know the geology of Georgian Bay is to know ourselves and our true place in this world. We are both, non-native and Native people alike, tenants of a region whose first geologic foundations were laid during a major plate tectonic collision almost 2,000,000,000 (2 billion) years ago during the Precambrian. At that time, our continent was much smaller in size. In fact, what is now eastern North America – the area from Sudbury to Boston — had not yet formed. The hard white quartzites of the Killarney area were deposited as sediments along the coast of the ancient continent and were buried, folded, and uplifted to form mountains that were not unlike the modern-day Rocky Mountains. This occurred some 1.8 billion years ago when a large landmass collided with and welded on to the southern margin of North America creating a crush zone called a "fold and thrust belt." The effects of the Penokean Orogeny are seen today as long white ridges of the La Cloche Range through which Highway 6 now meanders south of Espanola toward Manitoulin Island. Between 1.75 and 1.45 billion years ago these rocks were invaded from below by granites that together make up the distinct red "sugarloaf" hills and the white ridges of Killarney.

During the subsequent Grenville Orogeny which began just after 1.4 billion years ago, Georgian Bay found itself once more at the very epicentre of yet another enormous tectonic collision. Part of South America bore down on North America and a 500 kilometre wide strip of new land extending from Sudbury to Boston, and from Labrador to Texas, and called the "Grenville Province" was welded onto the ancestral North American continent and the Grenville Mountains formed where Georgian Bay now is. This globally significant event saw the formation of the massive supercontinent called Rodinia. By 800 million years ago, the towering Grenville Mountains were no more having been eroded to create the low-relief and distinctively knobbly surface of the Canadian Shield we know today. As such, this now iconic landscape is one of the oldest on the planet.

After the breakup of Rodinia some 700 million years ago, the newly formed Iapetus Ocean flooded across the flat-lying Canadian Shield which was then situated at the equator. The warm clear seas between 600 and 300 million years ago left thick layers of pale-coloured limestone and darker shales on the Shield below. These are called "Paleozoic cover rocks" in contrast to the "Precambrian basement" below. These sedimentary rocks are rich in fossils that reflect an amazing diversity of ancient marine life. Where these sedimentary rocks are exposed along the length of the Niagara Escarpment, on the west side of Georgian Bay northward along the Bruce Peninsula and on to Manitoulin Island, they commonly form bare limestone plains.

There is a huge and as yet unfilled gap in our knowledge of the history of Georgian Bay after 300 million years ago. Much rock has been scraped off by enormous ice sheets which began to form over northern Canada during successive ice ages starting 2 million years ago. The Great Lakes basins, including Georgian Bay, were dug out and deepened by successive North American ice sheets that were as much as 3 kilometres thick – that's 6 times the height of the CN Tower. The last great glacier only left the Georgian Bay area about 11,500 years ago when Georgian Bay was briefly flooded under the cold iceberg-laden waters of glacial Lake Algonquin, the much larger

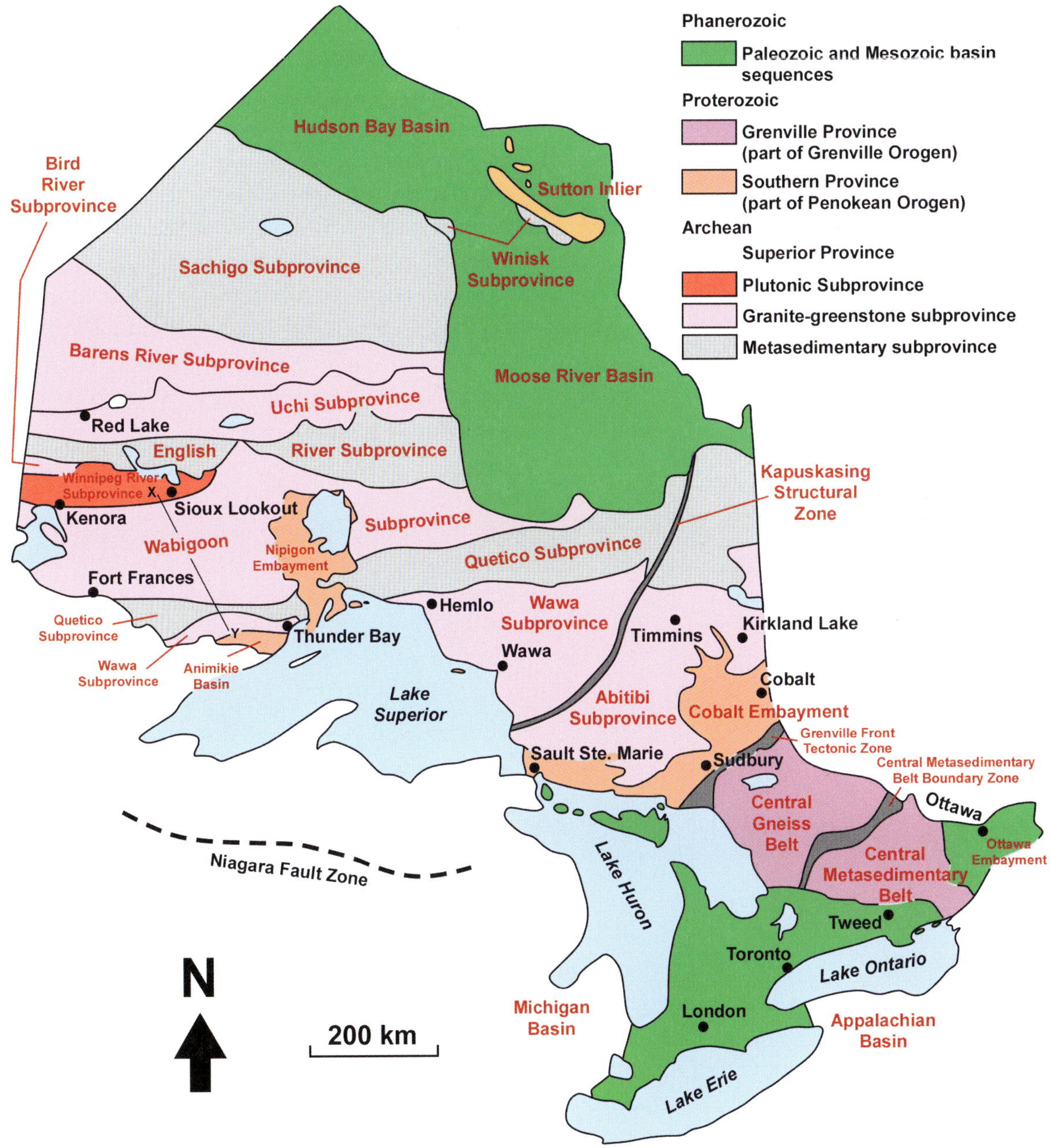

***Ontario's geology** consists of the Canadian Shield, and is made up of numerous provinces and subprovinces which took 2 billion years to assemble, and overlying younger cover rocks deposited in the last 600 million years. Those cover rocks in southern Ontario, adjacent to Georgian Bay, span the time period from about 500 to 350 million years. Shield rocks continue south under the younger cover rocks, reaching as far as Texas and New Mexico. Source: Ontario Geological Survey.*

ancestor of modern Lake Huron and Georgian Bay whose shorelines rim the modern Bay. The final retreat of the ice sheet opened the door for northward migrating humans who had travelled from Asia several generations earlier. Plants and animals also followed this northward migration. Abrupt and dramatic changes in lake levels occurred over the last 10,000 years. In fact, between about 8,000 and 5,000 years ago the Great Lakes were not only much smaller but also disconnected from each other. Furthermore, much of the floor of Georgian Bay was dry and forested in a warmer, much drier climate. Lake levels have risen since due to wetter cooler climates that have prevailed after 4,000 years ago.

Much of Georgian Bay is on the move, parts of it rising upwards by as much as 20 centimetres a century as the Earth's crust continues to rebound from the weight of the last ice sheet. At the same time, Georgian Bay (and the North American plate of which it is part), is moving inexorably westward by 4 centimetres a year driven by widening of the North Atlantic Ocean.

Thanks to plate tectonic drift and glacial rebound, this beautiful place we call Georgian Bay is on the move. It is not where it was yesterday — and it will be in a different place tomorrow.

Back to the Precambrian

By the late nineteenth century, geologists widely recognized the vast age of our planet and the idea of "deep," largely unexplored time. That said, much remained to be learned of the history of our planet that was still popularly thought to be only 6,000 years old and formed in less than a week. The rocks along the northern and eastern shores of Georgian Bay were particularly challenging. Not only were they complex and lacked fossils (a good indicator of age), but the amount of time they recorded seemed beyond comprehension, perhaps unmeasurable. However, the discovery of radioactivity in 1896 triggered a seismic shift in geology. Before this the highly influential views of Lord Kelvin held sway. His model showed a hot early Earth no more than 40 million years old that was cooling rapidly. But with new

The Birthplace of Plate Tectonics

The ancient Shield rocks exposed east of Georgian Bay are forever linked with John (Jack) Tuzo Wilson, the Canadian pioneer of plate tectonics in the mid 1960's; a new paradigm that explained how our planet works, where some two dozen large rigid plates move across the surface of the Earth over softer rock below. Wilson identified those places where plates are forming today (along mid-ocean ridges) and those places where some are being destroyed (in subduction zones), in the process continuously re-configuring the continents and oceans. Wilson realized that continents such as North America, grow with time as a result of collisions with other land masses. Using air photographs to see the broader landscape features missed by other studies he identified abrupt geologic boundaries within the Canadian Shield which he argued recorded its progressive enlargement over billions of years by accretion of new crust. Being able to age date rocks was key and he established a laboratory at the University of Toronto to test his ideas. Today after painstaking work in Georgian Bay and its many islands by geologists such as Nick Culshaw of Dalhousie University, Mike Easton at the Ontario Geological Survey and the late Tom Krogh at the University of Toronto, the geology of that part of the Shield exposed between the French River in the north, Honey Harbour to the south and eastwards through Muskoka towards Ottawa is now recognized as one of the world's best records of the accretion of smaller platelets (now preserved as "terranes" and "subprovinces") to create a bigger ancestral North America during the Grenville Orogeny. The rock left behind (collectively known as the Grenville Province which extends from Mexico to Labrador) was the last piece to be added to the Canadian Shield a billion years ago. The deformed gneisses of Georgian Bay record deformation of hot soft rock at near the base of plates; clues to how rigid plates are able to move across the Earth's surface.

Above all, Tuzo Wilson possessed the ability to see through all the detail assembled by previous generations of geologists and glimpse the bigger underlying truth about how the Canadian Shield (and all others world-wide) had been put together. Wilson died at age 84 in 1993. He and his family had a cottage on Go Home Bay ('Awry on the Rocks') and frequently sailed a Chinese junk (the 'Mandarin Duck') which he had imported from Hong Kong during one of his frequent visits to China. Much of our current global knowledge of the workings of ancient plate tectonics was conceived in Georgian Bay.

'To discover order and intelligence in scenes of apparent wildness and confusion is the pleasing task of the geological inquirer.' Thoughts on a Pebble.
G. A. Mantell 1849

No mortal mother made me
To feel the warm blood flow,
The years have not betrayed me
To hunger and to know.

A wiser mother made me;
From searings and from shock
The cooling years have stayed me.
I am the rock.

—The Rock
Barker Fairley (1887-1986)

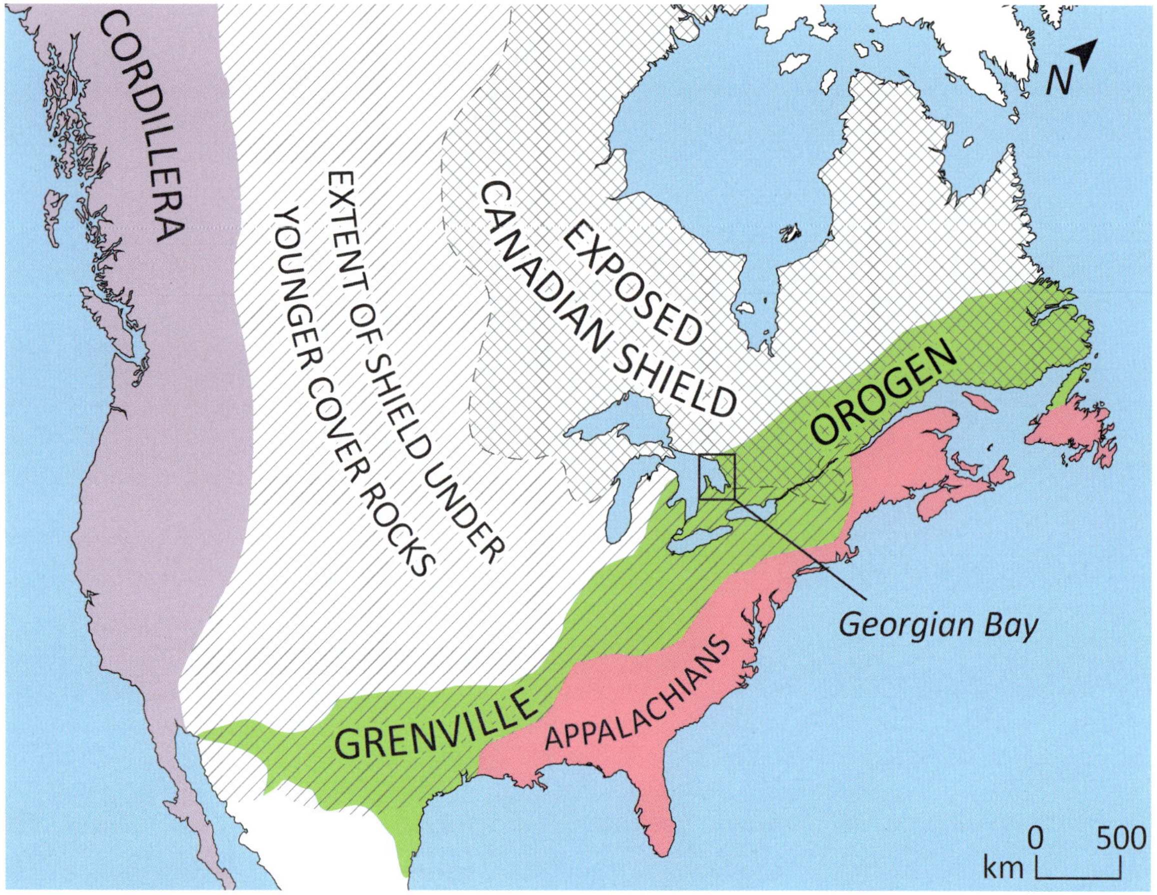

The Canadian Shield as we know it is merely the exposed part of a much larger expanse of Precambrian rock now buried by younger sedimentary rocks. In fact, the Shield underlies a substantial portion of North America. Its oldest part (about 4 billion years) is found in the far north (near Yellowknife) and the rocks of the Grenville Orogen record the last piece of the Shield to be added about 1 billion years ago. The rocks of the Cordillera and Appalachians and much of Maritime Canada, are more recent additions to the North American continent and were added around the periphery of the Shield. It is worth emphasizing that the Great Lakes are geologically very recent in age and fill basins excavated by ice sheets over the last 2.5 million years.

understanding of radioactivity it was discovered that the Earth had an internal heat source that had kept the planet warm for hundreds of millions of years and would do so far into the future.

The term Grenville has long been attached to the Shield rocks along the northern and eastern margins of Georgian Bay. The term "Grenville Series" had been introduced by Sir William Logan in 1863 from his observations of the complexly banded metamorphic rocks (called gneiss) near Grenville Township along the north shore of the Ottawa River. Fairly soon it was evident that the Grenville rocks stretched as a broad belt westward from Ottawa to the shores of Georgian Bay. Some thought the rocks to be the original primeval molten crust of the planet while others argued that the banding in such rocks had been formed by running water (the so-called "Neptunist" school of geology). Slowly, it became apparent that they contained vestiges of even older rocks that had been profoundly changed by deep burial and heat (metamorphism). Logan wrote in 1857 that "it leaves no doubt in my mind that they are a series of very ancient sedimentary deposits in an altered condition." In the end it would take almost a century after Logan's first observations for geologists to arrive at the fact that these rocks were more than a billion years old and recorded an ancient orogenic belt marking the growth of an early North American continent.

The United Plates of Ontario— The ancient supercontinent of Rodinia

The geology of Georgian Bay is often difficult to comprehend at ground level. In fact, much of the story is best revealed from space. The differing hardness and resistance to erosion of the many varied rocks that make up the Grenville Province is reflected in the very uneven and often distinctly corrugated surface of the Shield. Apparently chaotic, the Shield shows a distinct pattern of long winding rock ridges and water-filled depressions. Satellite images and aerial mapping of the magnetic properties also reveal great sweeping swirls in the rock and giant folds tens of kilometres across that form a distinct regional "grain" much like wood. This striking pattern is reflected in the distribution and shape of lakes which closely follow the underlying bedrock structure. The giant folds are the deformed mantle roots of the Grenville Mountains now laid bare at the Earth's surface by deep erosion. They were created deep underground when tectonic "platelets" were thrust against each other by colossal forces to form an enormous tectonic mosaic. Each piece of the mosaic is given the name "terrane" such as the Go Home, Seguin, Rousseau, Shawanaga, Moon River, Muskoka, and Britt. These terranes are all of different ages having had their own geologic story long before being pushed together like tiles on a roof. Some were far-travelled islands made of volcanoes (like modern-day Japan), some were little micro-continents carrying sedimentary rocks while others were slabs of ocean floor made of basalt. Nevertheless, all were slammed together during the Grenville Orogeny about 1 billion years ago to create a sort of "United Plates of Ontario" (now called the Grenville Province). Just as the humans that were eventually to live on these rocks have all come from someplace else, so too have its rocks.

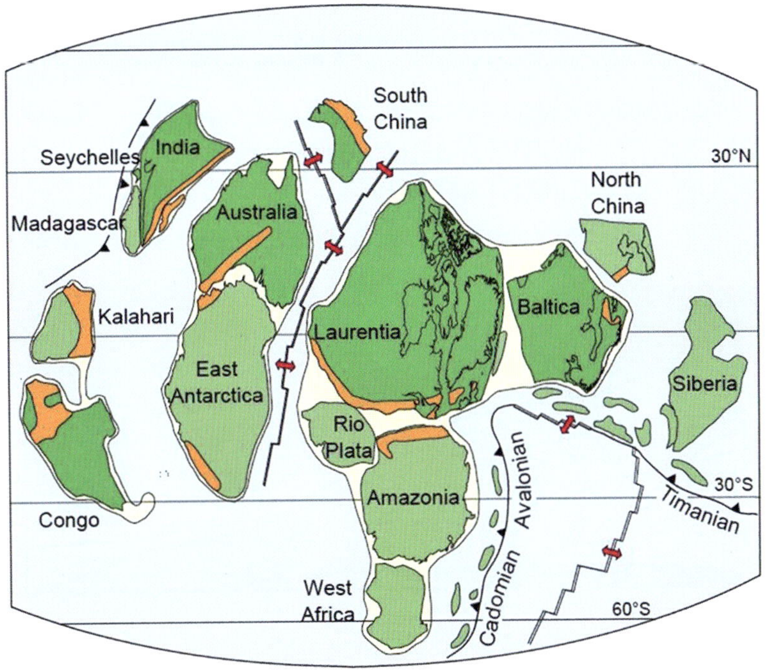

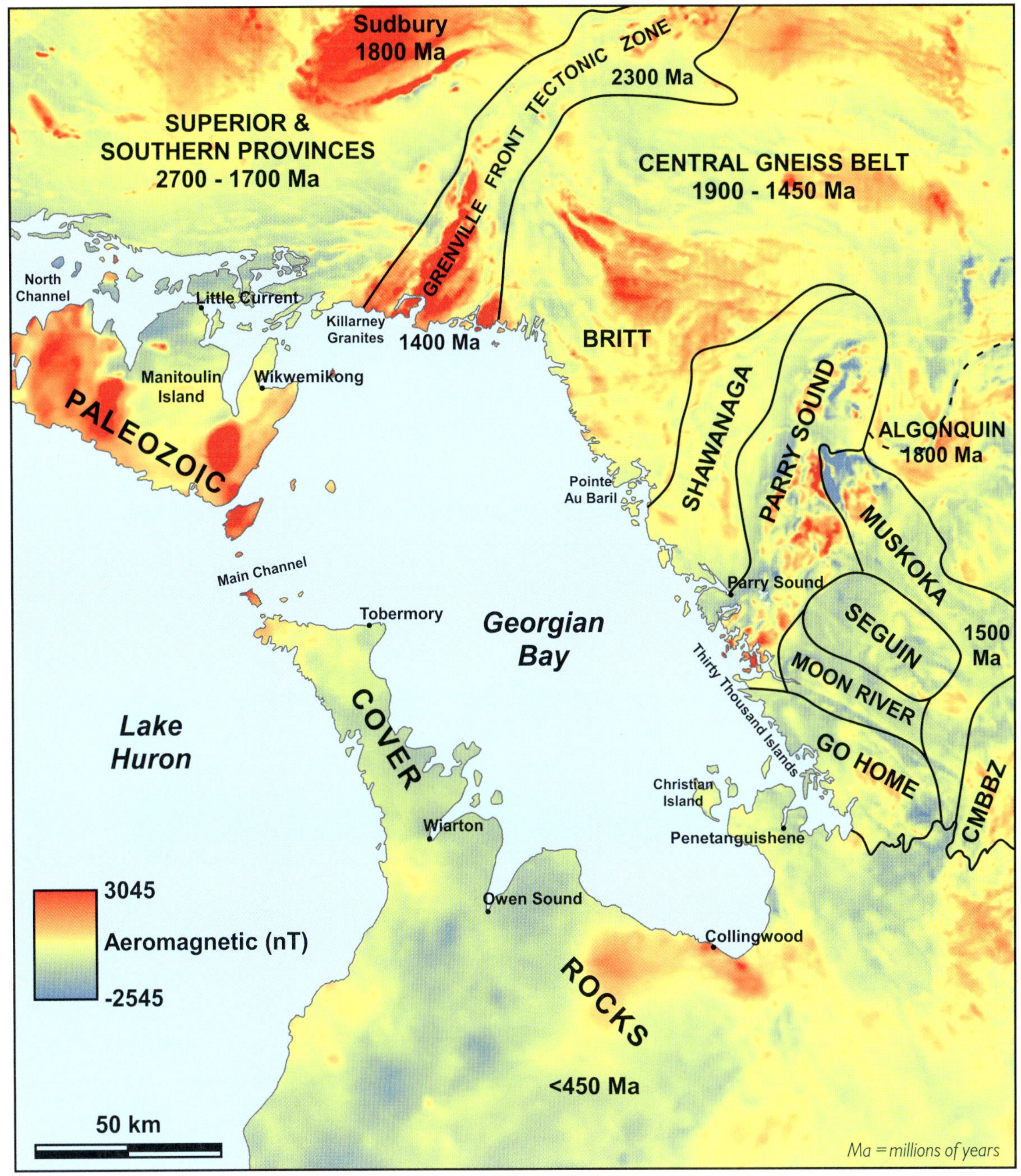

Top: **Using a wide array of techniques,** *geologists have solved the mystery of what planet Earth looked like at the time of the Grenville Orogeny between 1.5 and 1 billion years ago. It was a time of the supercontinent Rodinia when Amazonia (the northern half of the modern continent we call South America) was conjoined with ancestral North America (Laurentia). This created a collision zone in which rocks were thrust onto the North American continent during the Grenville Orogeny. The area we now call Georgian Bay lay deep below the Grenville Mountains only to be exposed much later when erosion lowered the mountains to expose their roots.*

Right: **Georgian Bay** *lies along the abrupt boundary of two completely different geologic realms: the billion-year-old hard crystalline rocks of the Canadian Shield to the north and east, and the much younger softer sedimentary rocks to the south and west. Shield rocks are a tapestry of "provinces" (e.g. Superior), "belts" (Central Gneiss Belt), and "terranes" (Parry Sound, Seguin, Shawanaga, etc.) which were crustal blocks and micro-continents in their own right before being pushed together by plate tectonic collisions to build a bigger North American continent. High mountains created by these collisions were worn down by 600 million years ago to expose their deep crustal roots and form the generally low relief and barren surface of the Canadian Shield. Warm shallow seas flooded across this surface after 500 million years ago to leave thick deposits of limestones, shales, and sandstones now covered by thick glacial sediments which underlie the rich farmland of the south. The magnetic properties of the rocks, when mapped from the air, and expressed here as different colours, reveal the complexity of the Canadian Shield that was deformed when rocks now at surface lay as much as 25 kilometres underground. Most of the rocks in the Shield are high-grade metamorphic rocks such as gneiss that are characterized by strong banding. Slow creep of these deep rocks when hot and soft allowed movement of overlying tectonic plates. Large areas of granite across the Shield reflect intrusion of magma at depth.*

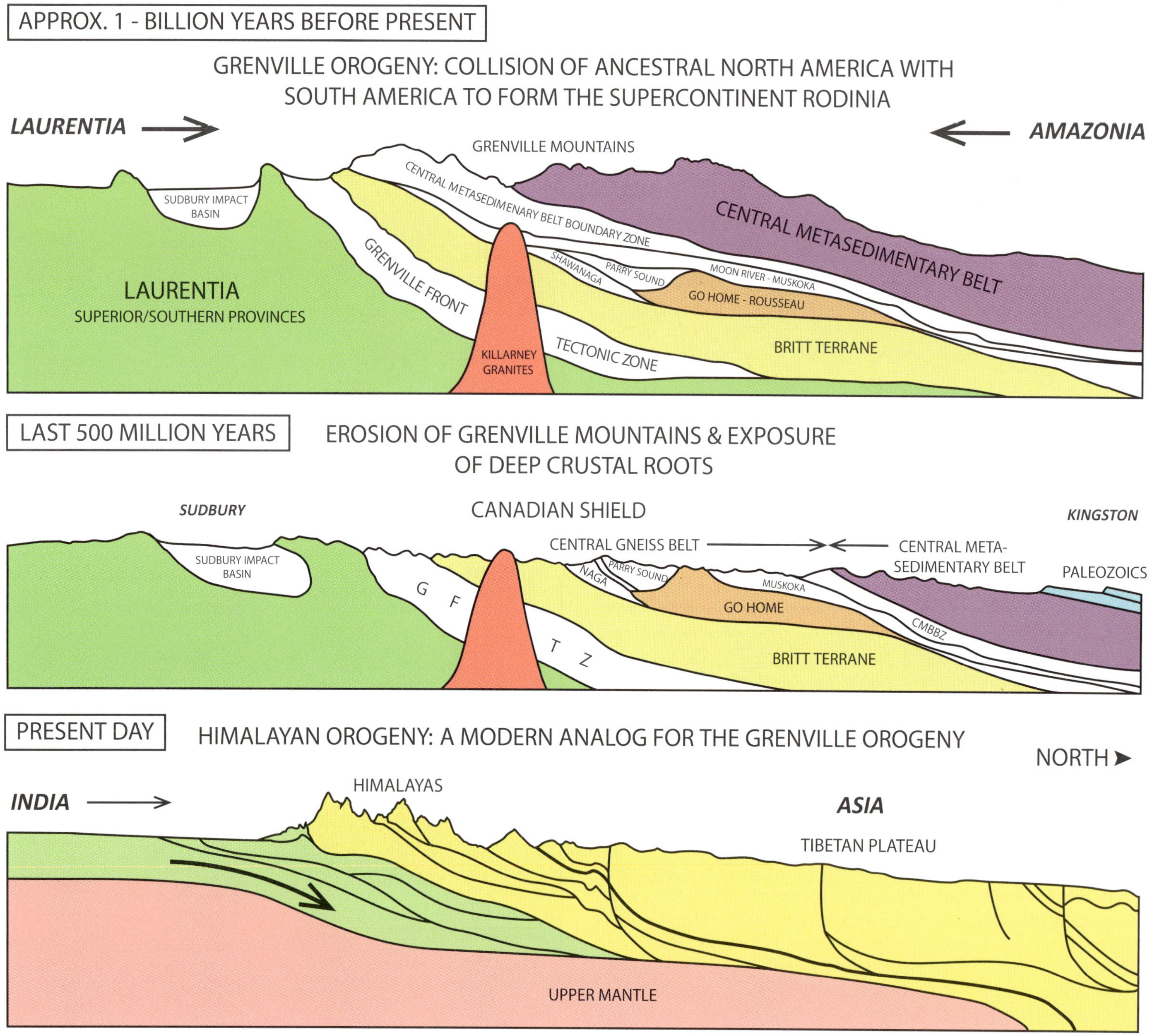

North-south cross-sections *through the Shield of eastern Georgian Bay show the basic structure consisting of imbricated crustal slabs akin to tiles on a roof and their evolution through the last billion years. These slabs or terranes were tectonic platelets that were pushed together and stacked one on another when ancestral North America collided with the northern part of South America. The situation in the Himalayas today is a good modern analog; here the Indian sub-continent is being rammed northward into Asia, pushing up the high Himalayan Mountains.*

Left: ***The Grenville Mountains*** *once dominated the early North American continent in a broad belt running from Texas to Labrador about a billion years ago. The scene today near Mount Everest in the Himalayas is similar to that in Georgian Bay at the time. About 500 million ago, the high mountains had been torn down to an almost level plain to reveal their deep crustal roots.*

Right: ***The deep roots of the Grenville Mountains*** *now laid bare by millions of years of erosion. The flat, low topography of much of the Canadian Shield of eastern Georgian Bay was established at least 500 million years ago just before it was covered with Paleozoic sedimentary rocks in warm tropical seas. These cover rocks have been stripped off, and the exhumed and much fractured Shield surface scoured by ice sheets in past ice ages. Now the region is flooded by Georgian Bay waters.*

During the Grenville Orogeny numerous terranes were pushed to the northwest, up and onto the margin of ancestral North America. This gives rise to the characteristic eastward "dip" (tilt) of the layers in rocks seen in the road cuts along Highway 400 or on the islands out in the Bay. The orogeny also gave rise to the regional "imbrication" of rocks where the youngest lie to the southeast toward Kingston and the oldest to the northwest near Killarney. In the process the original rocks were heated and almost melted (metamorphosed) beyond recognition to produce the banded gneisses that typify the region. These are some of the most spectacular looking rocks found anywhere in the world and are a magnet for geologists and artists alike. Formed at temperatures as high as 800° C and pressures up to 6,000 times atmospheric pressures, these rocks tell us about the processes that operated some 25 kilometres deep at the base of moving tectonic plates. Now exposed at surface, the metamorphic rocks of Georgian Bay give geologists a unique lesson into how rigid crustal plates are able to glide across the surface of the planet.

Rocks of the Thirty Thousand Islands

The rocks of Georgian Bay can be divided into two major groups. To the west are the fossiliferous sedimentary rocks of the Bruce Peninsula and Manitoulin Island which are about 425 million years old. These include limestones dominated by the mineral calcite, and which contain numerous fossils of marine organisms, and dolostones where calcite has been replaced by the much harder mineral dolomite (often destroying fossils in the process), together with grey or red coloured shales. These rocks contrast with the much older (a billion years plus) and harder Precambrian metamorphic and igneous crystalline rocks of the Shield to the east.

The most common metamorphic rock in eastern Georgian Bay is gneiss (a German word pronounced "nice") which is easily identified by its very distinctive layering (called foliation). These are often arranged in spectacular folds. Gneisses are the lubricant of plate tectonics; their deformation allows the cooler, more brittle crust to move overhead as a solid plate several tens of kilometres thick. Their typical banded appearance is the result of the growth of minerals such

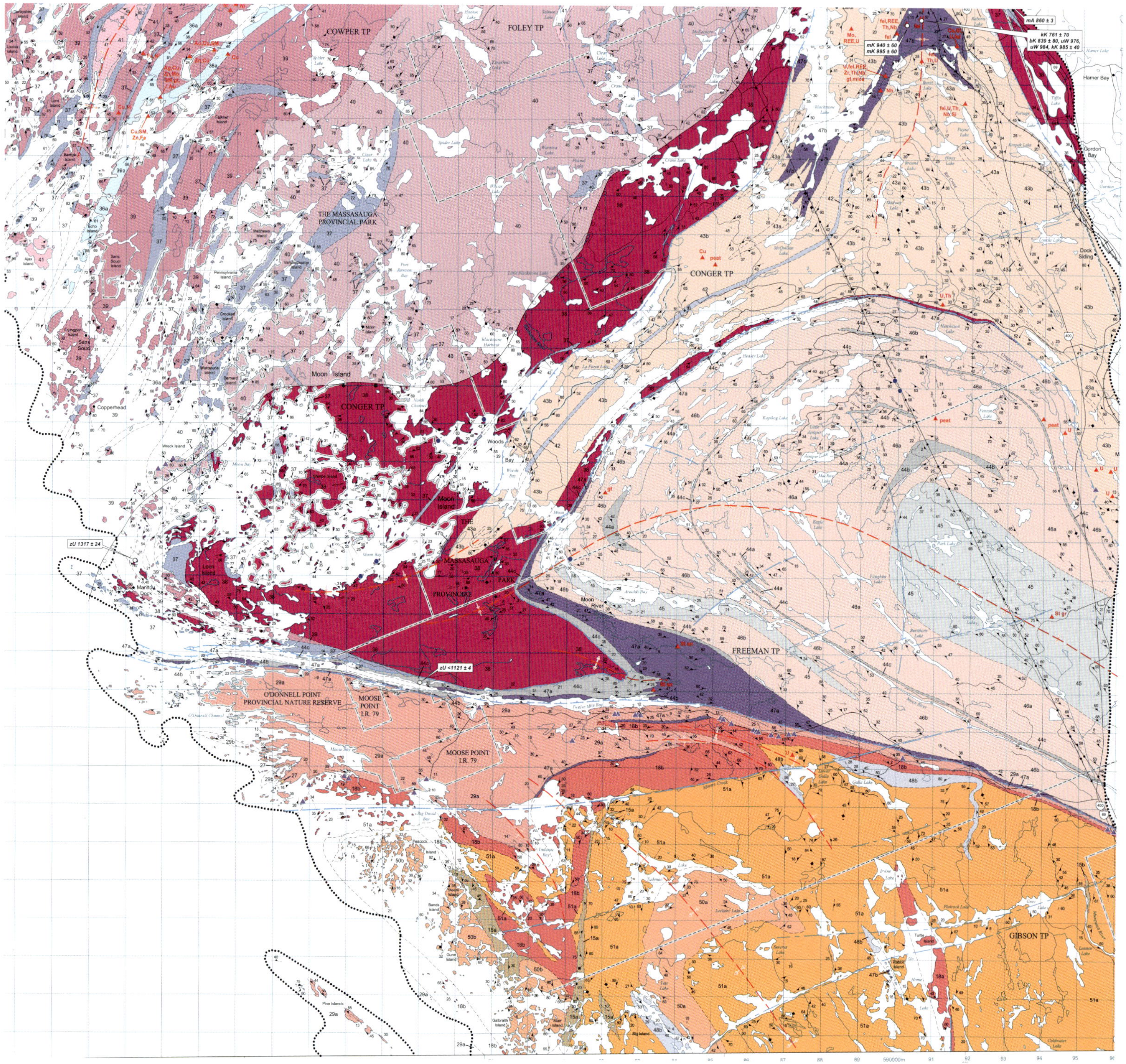

"Beneath all the wealth of detail in a geologic map lies an elegant orderly simplicity."

J.T. Wilson

Complex regional folding *of the gneissic rocks of the Moon River, Go Home and Seguin terranes is revealed by detailed mapping by Nick Culshaw and colleagues of the Ontario Geological Survey. The long west-east indentation of 12 Mile Bay follows the boundary of the Moon River and Go Home terranes. Courtesy of Ontario Geological Survey.*

Black 'mafic' dykes made of igneous rock (typically amphibolite rich in iron and magnesium) were forced at high temperature and pressure into lighter coloured granulites.

Banded gneisses

as quartz, feldspar and hornblende in distinct layers under very high pressures and temperatures. These layers are then repeatedly folded and re-folded as the hot rock is squeezed and deformed, just like hot toffee, below the base of the moving plate.

Local melting under intense pressures created another type of more completely metamorphosed rock called migmatite where layers have a much more irregular appearance.

Coarse-grained gabbros and the finer grained equivalents called basalts are the product of magmas derived from the upper mantle. These igneous rocks have commonly been metamorphosed to form dark, grey-green coloured amphibolites. An unusual form of gabbro is anorthosite which is common in the Parry Sound area. This unique rock is dominated by white feldspar. Such rocks are common on our moon where they form the pale coloured lunar highlands.

Pale-coloured marbles are identified by their granulated sugar-like crystalline appearance. This is the result of burial and heating of seafloor limestones pushed kilometres underground during the Grenville Orogeny. They are the softest rocks anywhere on the Shield and are associated with lakes and valleys where erosion has selectively removed them.

Pinkish-coloured granite (from the Latin *granum*, meaning *grains*) is a most common intrusive igneous rock found across the Shield. It is light orange coloured because it is dominated by silica and feldspar minerals and it is easily distinguished from gneiss because it lacks any banding. Granite typically forms broad dome-like hills (plutons) and has often been metamorphosed to form a weakly banded pink rock called granulite. This is another very common rock found around and north of Parry Sound. Look for an obvious swirly structure that lacks the distinctive banding of gneiss.

Precambrian sedimentary rocks occur along the northern rim of the Bay. The most noticeable are the white quartzites of the La Cloche Range north of Manitoulin Island. Many different types are found across Georgian Bay as erratic boulders that were torn up and brought south by glaciers during the last ice ages. Particularly distinctive types are the "puddingstones" and the distinctive quartzites that contain blood-red chips of jasper commonly found scattered along the shores of the North Channel.

The eastern coast of Georgian Bay and Thirty Thousand Islands lies within that part of the Grenville Province called the Central

Classic banded gneisses *of eastern Georgian Bay. These are composed of alternating bands of dark coloured amphibolite derived from the metamorphism of basaltic magma containing olivine and magnesium, and lighter coloured bands of a granitic composition dominated by feldspar and quartz.*

Folded gneiss seen on the Pancake Islands at the mouth of Parry Sound.

Gneiss Belt (CGB) because of its predominance of gneiss. The belt is about 150 kilometres wide and in sharp contrast with the Central Metasedimentary Belt (CMB) which is on its eastern margin toward Ottawa. The CMB, however, is made up of metamorphosed sediments deposited in seaways that closed as surrounding landmasses were pushed into each other and fused together during the Grenville Orogeny. The narrow zone between the two belts is known as the Central Metasedimentary Belt Boundary Zone (CMBBZ) which continues southward under the cover of younger sedimentary rocks that pass under the north shore of Lake Ontario, near Pickering before crossing the lake, the Niagara Peninsula, and then into Pennsylvania. This was a major thrust fault in Precambrian times and even today is associated with small to moderate earthquakes with the potential for larger events. We call these "intraplate earthquakes" because they occur near the middle of the North American plate unlike the more

***Some gneissic rocks have a spotted appearance** created by the growth of isolated crystals of pink-coloured feldspar (bottom) and paler varieties (top). These were rotated as the parent rock was stretched and deformed under high pressures and temperatures. This provides geologists with valuable information about the direction plates travelled.*

***Intense deformational forces** at work deep below the Grenville Mountains a billion years ago is clearly evident in these complexly folded gneisses exposed in road cuts along Highway 400 at Parry Sound. Orange-coloured dikes have cut through later. These rocks are part of what is called the Parry Sound Shear Zone.*

Granulites *are metamorphosed granites: note the 'swirly' structure very different from the parent massive granite, and banded gneisses.*

Opposite page: ***The most intensely deformed rocks of Georgian Bay*** *are mylonites seen here in a roadside outcrop. The individual layers are so thin and perfectly parallel that they resemble pages in a book.*
Right: ***This is migmatite*** *and results from the partial melting of gneiss giving rise to a "blobby" appearance where the classic banding typical of gneisses has been destroyed.*

common earthquakes that occur at its margin, notably along the western coast of North America.

The northern limit of the Grenville Province lies immediately north of Georgian Bay along a sharp boundary called the Grenville Front Tectonic Zone (GFTZ). This zone occurs just south of Sudbury. Very old rocks (over 2 billion years) lie to the north, including the great Sudbury impact structure which was formed 1.8 billion years ago when the Earth was struck by a 10 kilometre wide meteorite.

The boundaries between each terrane within the CGB took the full brunt of the Grenville collision between 1,200 and 1,000 million years ago. This created narrow corridors of intensely metamorphosed and stretched rocks called shear zones. The Parry Sound shear zone underlies the community of Parry Sound and extends south to Parry Island and out across the many islands of Georgian Bay near Sans Souci. Twelve Mile Bay is eroded along another prominent shear zone between the Moon River and Parry Sound terranes where its spectacular folded rocks have been the principal focus of artist Ed Bartram whose island lies astride the zone. These shear zones are traversed by Highway 400 and are composed of beautifully laminated mylonites and the more thickly banded gneisses. Their complex folds mimic the larger regional structures and tell us the rock has been stirred by giant subterranean forces. Huge masses of deeper underlying iron-rich mantle rocks such as black coloured gabbro and amphibolite were incorporated into the shear zone forming large dark inclusions in the generally light coloured gneisses.

The end of Rodinia

Supercontinents are temporary tectonic constructions and are eventually torn apart (they are said to "rift") by internal forces that they themselves put into play. By covering such a large part of the Earth's surface they restrict the escape of heat from the mantle. Mantle rocks are dry and have a consistency much like hot toffee but when warmed by an overlying supercontinent they slowly rise to the surface as giant plumes (think of that lava lamp in the basement). Relieved of the enormous pressures that keep them from melting at depth, the tips of the plumes undergo "decompression melt" which is the releasing

The stretching process *that took place deep below the Grenville Mountains is clearly evident here. The dark coloured rock is amphibolite derived from the metamorphism of gabbro and basalt, and alternates with more lighter colored bands dominated by quartz and feldspar.*

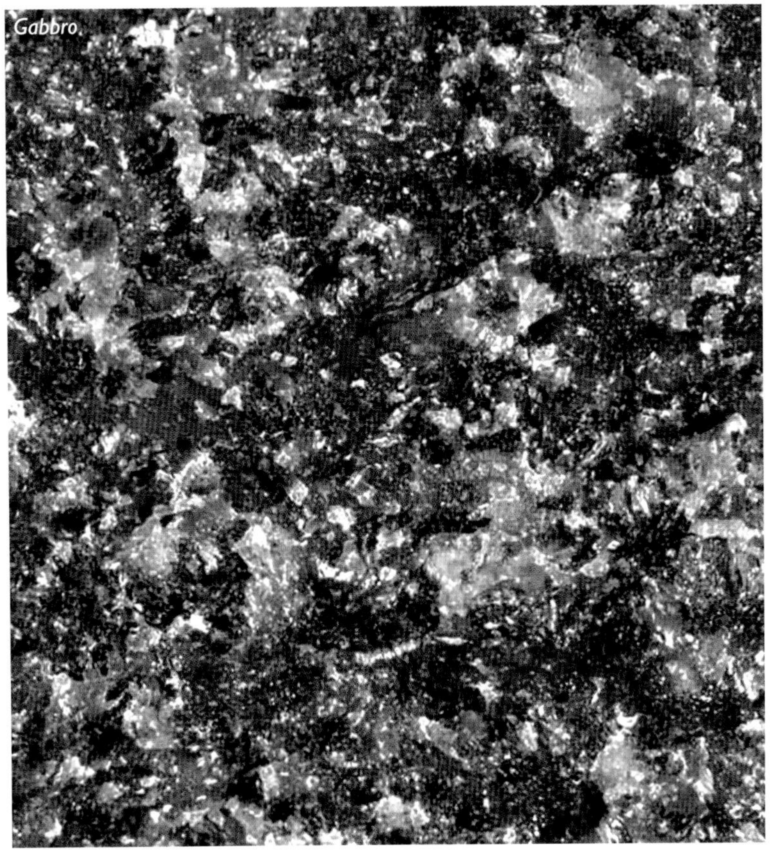

of molten magmas that force their way upward to the Earth's surface. The many intrusive dikes seen along road cuts and as dark coloured ribbons snaking their way across the islands in Georgian Bay record this episode as Rodinia was stretched and eventually broke up. The key here is that it didn't break up along the original boundaries of Laurentia (near Sudbury), but far to the south so that the new coastline of Laurentia lay near what is now Atlanta where it faced a brand new Iapetus Ocean. At the same time the mighty Grenville Mountains were also being demolished by rivers and enormous rock falls to be replaced by a low relief plain we now call the Canadian Shield. This was the surface that was to be flooded by the encroaching waters of the Iapetus Ocean that ushered in a new chapter in the story of Georgian Bay's history where marine life flourished in shallow, warm seas.

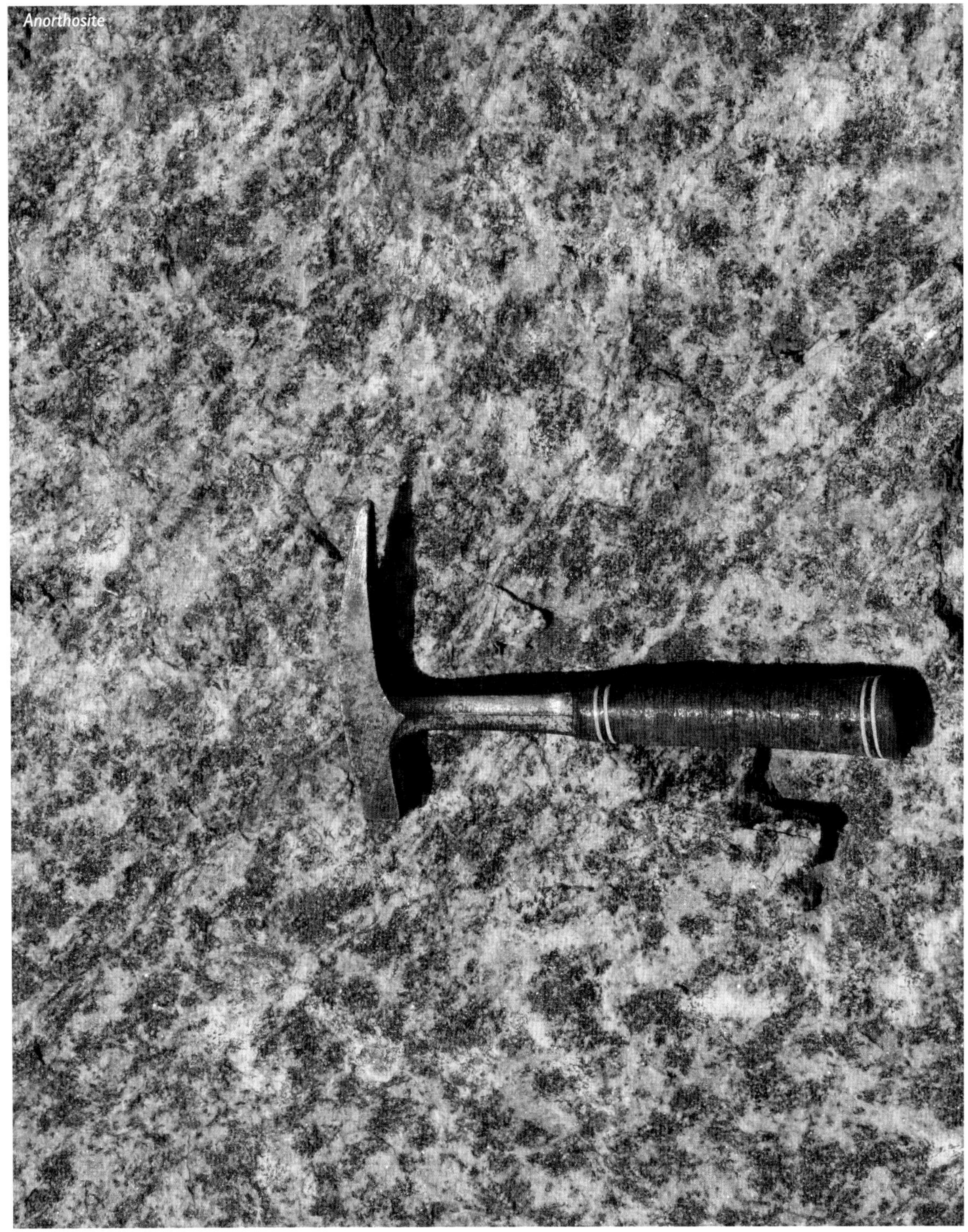

***Gabbro** (top) is coarse grained and an unusual form is **anorthosite** (right) which is dominated by white feldspar and is common in the Parry Sound area. Such rocks are common on our moon where they form the pale coloured lunar highlands.*

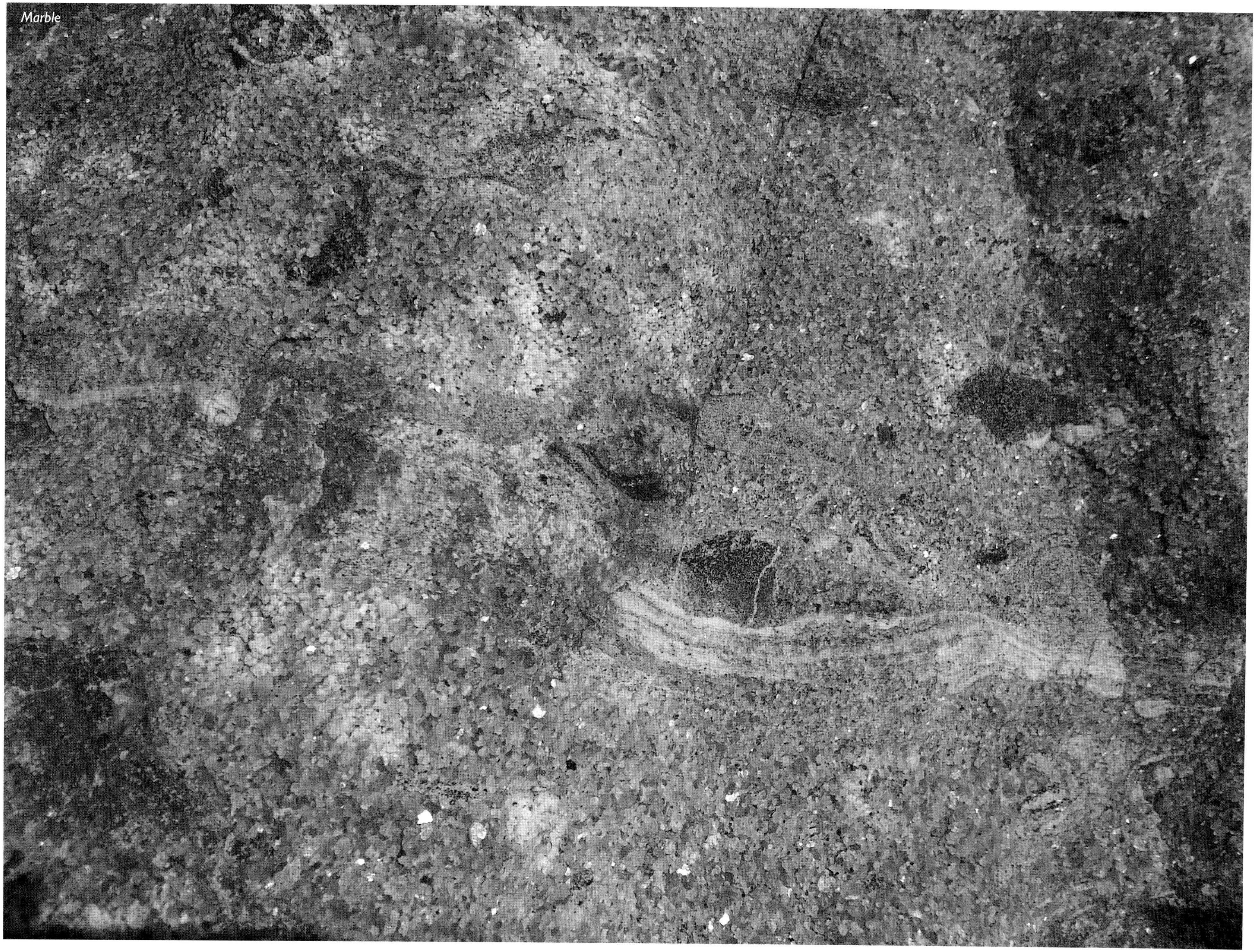
Marble

Granite

Geologist Tom Krogh *working on a map of the area around Twelve Mile Bay in eastern Georgian Bay showing the complex regional folding that has affected rocks of the Canadian Shield.*

The Paleozoic world and the birth and death of Pangea

For much of its history after Rodinia, Georgian Bay was part of Michigan – the Michigan Basin that is. This was the beginning of the quietest chapter in North America's entire geologic history. Freed from the clutches of Rodinia the continent became surrounded by warm, clear oceans where marine life proliferated and took on radically new forms. Protective skeletons such as shells protected soft bodies and allowed organisms to burrow into seafloor sediments in search of food and cover. North America was flooded by shallow seas that stretched from one end of the continent to the other leaving thick layers of rocks called limestones composed of broken shells. Most of this cover has been entirely stripped off the Shield but vestiges of it still survive in eastern Georgian Bay on the Limestone Islands.

Vertical bands *(called dikes) of granite are a ubiquitous feature of the Shield and result from upward injection of magma from sources deep below. They helped stitch together the many crustal blocks that had been pushed together during the Grenville Orogeny. Where magma cooled very slowly (on right), the dikes are characterized by large crystals of feldspar and quartz. These are known as pegmatite dikes and often show distinct zones of crystals.*

By 440 million years ago, the quiet warm world of limestones was coming to an end with eastern North America once again in collision with other landmasses, this time Africa and parts of what is now Scandinavia. This was the beginning of the supercontinent called Pangea. Ocean floor crust was pushed down below North America (a process called subduction) while volcanoes grew upward spewing ash into the interior. Large mountains (the Appalachians) dominated the skyline of eastern North America and large rivers carried muddy and sandy sediment westward into Ontario. The day of the limestones was largely replaced by shales and sandstones.

The Niagara Escarpment is made up of a rock type called dolostone which is the result of the chemical alteration of limestones when deeply buried. The escarpment is the outcropping edge or "rim" of a dolostone layer and creates the backbone of the Bruce Peninsula and Manitoulin Island, and the steep west coast of Georgian Bay. It can be traced around the Michigan Basin. Low relief limestone and dolostone plains devoid of soil are exposed on Manitoulin Island and across most of the Bruce Peninsula. Glacial erosion and deposition during successive ice ages has greatly affected the Niagara Escarpment by cutting deep valleys such as Owen Sound and the prominent deep-water channel that separates Manitoulin Island and the Bruce Peninsula.

The demise of Pangea began about 200 million years ago when eastern North America and the conjoined landmass of Europe underwent stretching prior to final rifting of the supercontinent and the creation of the early North Atlantic Ocean. The St. Lawrence and Ottawa Valleys are failed rifts where the crust stretched but did not break completely. The Ottawa Valley trends northwest and splits in two with one arm under Lake Nipissing and another under Lake Timiskaming. The Georgian Bay area was similarly affected and faults control the location of its principal rivers such as the French, Key, Magnetawan, and Shawanaga and many lakes and inlets such as Collins Inlet near Killarney. This explains why these tend to be

Blood red garnets *that resemble pomegranate seeds are a very common mineral in the metamorphic rocks of Georgian Bay. Garnets require very high pressures and temperatures to form and are typical of rocks that were pushed down to depths of at least 20 kilometres during the Grenville Orogeny.*

"Screams from the tortured banding of the minerals." *Gaile McGregor, 1985 from The Wacousta Syndrome: Explorations in the Canadian Landscape (University of Toronto Press). An apt description of the complexly folded gneisses seen at right.*

Alexander Murray (1810-1884) was a British naval officer made redundant at the end of the Napoleonic Wars in Europe. He immigrated to Canada and turned to geology as a profession after an unsuccessful stint as a farmer near Woodstock in 1837. He was a friend of Sir William Logan, the first director of the Geological Survey of Canada which was set up in Montreal in 1842. Though largely self-taught in geology, he was hired as a surveyor in Canada West (what is now approximately central and southern Ontario) which at the time was still heavily forested with few roads. Murray's work was used to attract settlers to the area. Murray is credited with recognizing that the Canadian Shield near Severn Bridge, just north of the Narrows at Lake Couchiching, is dominated by highly metamorphosed gneiss. He made extensive studies of the rocks of the Huronian Series (which he named) from Sault Ste. Marie to the North Channel of Lake Huron to Timiskaming. He was the first to identify the bituminous deposits in southwestern Ontario (in 1851) several years before the commercial development of oil (1858) at Petrolia. He recognized the mineral potential of the Sudbury area, specifically cobalt and nickel, and discovered the famous "kettle rocks" at Kettle Point near Ipperwash in Lake Huron. He was shipwrecked in a fierce squall in Georgian Bay in 1861 while undertaking the first geologic exploration of the fossil-rich limestones of the Bruce Peninsula north of Wiarton. He was also well aware of the use of fossils in correlating layers of rock from one part of southern Ontario to another and wrote in 1845: "Mr. Ross, who surveyed Barrie Township, gave me a specimen found in the 2nd Concession of Collingwood township north of the Blue Mountains, on the shore of Lake Huron. This specimen and the strata of Windsor are identical in their fossil content and mineralogy. Therefore one can conclude that the formation extends from one lake to the other. In the outcrop the rock is jet black shale. It is hard, fissile and breaks easily. It exfoliates into thin sheets and the surfaces of these sheets are covered with a very thin bituminous skin. Under weathering conditions the rock disintegrates rapidly and finally becomes a black shale that forms a good soil. Fossils described by American geologists as characteristic of the Utica shale in New York State are abundant in this rock. In Whitby I found specimens of the trilobite called Triarthrus beckii with shells of genera avicula and orthoceras."

Today Murray's contribution to the geology of the Georgian Bay region is almost forgotten as his reputation has been overshadowed by the work of Logan. He is better known instead for his later exploration and mapping of the interior of Newfoundland in the 1870s.

John Jeremiah Bigsby (1792-1881) was another soldier who made great contributions to our early understanding of the geology of Georgian Bay. He was a military physician and saw service in North America (1818-1826) when he was attached to the Boundary Commission charged with establishing the precise border between the USA and Upper Canada after the War of 1812. He made many observations of the geology of the Bruce Peninsula including the first reports of fossils in its limestones. His report on "The Geology and Geography of Lake Huron" presented to the Geological Society of London in 1824 was later heralded by Sir William Logan, then Director of the Geological Survey of Canada, as the "first essay of any importance upon the fossils of Canada." Bigsby published a classic of early North American travel and observations called the "Shoe and Canoe" in 1850 and was the first geologist to examine the Oak Ridges Moraine north of Toronto.

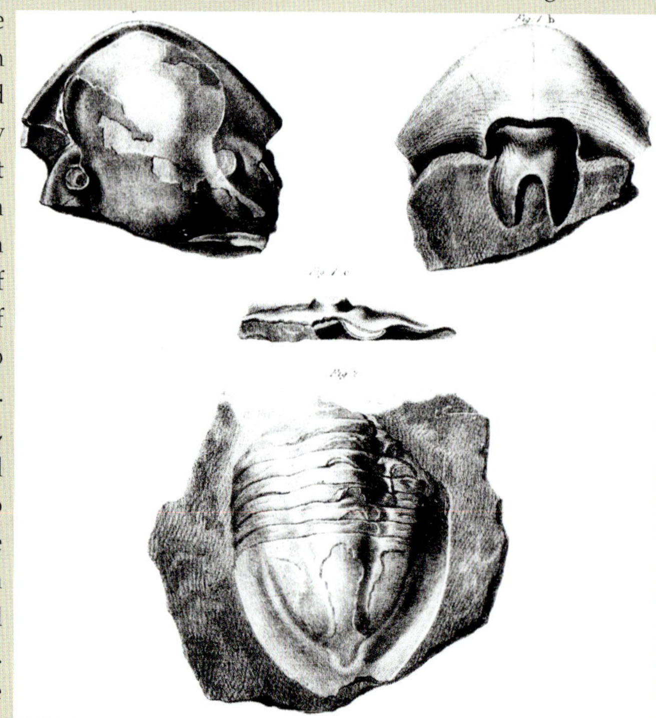

The Niagara Escarpment seen from the Cup and Saucer hiking trail on Manitoulin Island.

Typical limestone fossils of Georgian Bay

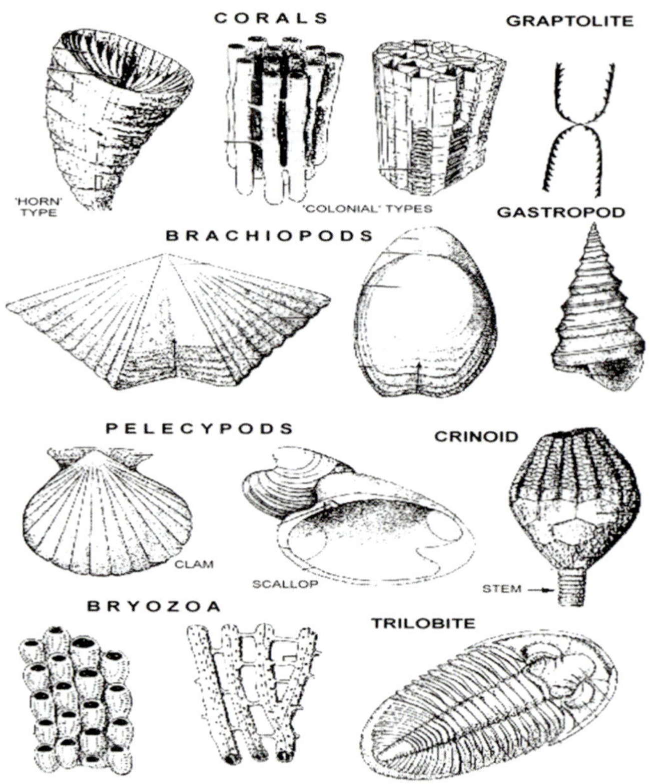

Crinoid

A limestone quarry at Wiarton on the Bruce Penninsula

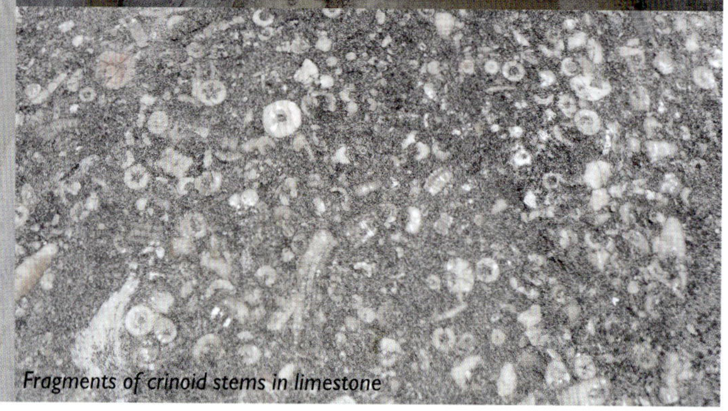
Fragments of crinoid stems in limestone

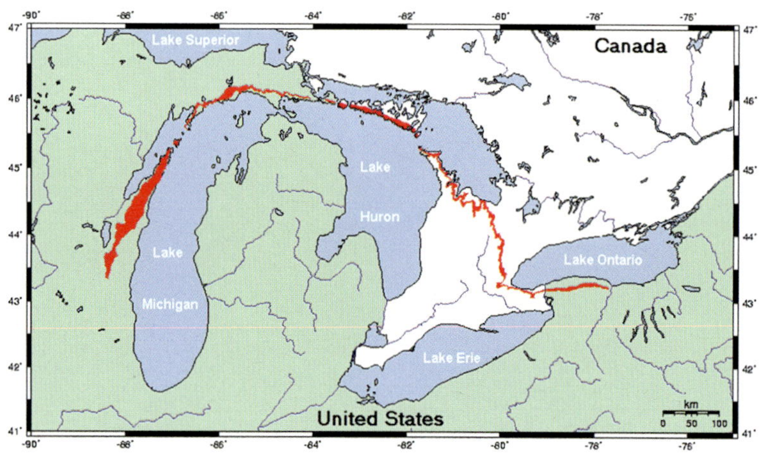

The flat-lying sedimentary rocks of the Bruce Peninsula and Manitoulin Island were deposited in shallow seas that flooded the Canadian Shield sometime after about 500 million years ago. These rocks are dominated by limestones made up almost entirely of the fossilized remains of a diverse range of organisms, and very commonly the broken stems of animals called crinoids ("sea lilies" or echinoderms) that grew on sea floor reefs when ancestral North America lay in the tropics.

very straight much like canals, and occupy narrow steep-sided bedrock valleys that often form a criss-cross pattern where one valley intersects another. Many rivers and lakes also follow the boundaries of the smaller crustal blocks that make up the Shield. This distinctive drainage system is said to be "structurally controlled."

Turning hard rocks into mush: when Georgian Bay looked like Africa

Over the past 50 million years, the rocks of Georgian Bay experienced deep weathering in a seasonally wet, warm climate. For much of this time even the Arctic had dense stands of giant redwoods. The hard gneisses of the Canadian Shield were weathered and reduced to a thick layer which, using modern-day central Africa as a model, would have been perhaps hundreds of metres deep, resting on unweathered rock below. The boundary between the two is known as a *weathering front*, while the weathered layer on top is called a regolith.

Some rocks such as marble were very prone to weathering. Other more resistant rock such as granite would have poked up

***The Limestone Islands** are the most easterly outcrop of Paleozoic sedimentary strata in eastern Georgian Bay. They can be found just north of the mouth of Parry Sound. They are remnants of the Paleozoic limestones that once covered much of the Canadian Shield. The remote and windswept islands lie some 30 kilometres offshore and are surrounded by dangerous reefs that can only be navigated when the lake is calm. Few trees are able to withstand the major storms that drive water (and ice) across the low, weather-beaten islands. These rocks record an ancient reef in which coral and other organisms flourished during the Ordovician period 450 to 500 million years ago. The Islands' limestone pavements display giant fossilized 460-million-year-old cephalopods which are a class of marine mollusks. Their current relatives include squid, nautilus, and octopus. Long glacially cut grooves record the flow of the last great ice sheet over Georgian Bay.*

through the regolith to form a high-standing landform called a *koppie,* so-named after those found today in central equatorial Africa (called kopjes ("little heads") in Afrikaans). These formations resemble huge piles of boulders stacked one on top of another. The scientific name for them is *monadnock* named after Mount Monadnock in the eastern USA. A monadnock is a residual hill resulting from weathering and lowering of the surrounding landscape. Many bedrock summits across the Shield would have resembled enormous piles of large rounded boulders forming a landform called a tor. Such boulders are simply unweathered blocks (*core stones*) freed by deep weathering along vertical and horizontal cracks. Weathered rock is washed out from the cracks and the core stones become perched on each other. Koppies and tors no longer look out across the Canadian Shield because they were bulldozed by ice sheets which simply scraped off the clayey weathered layer and moved core stones southward into Southern

On the western side of Georgian Bay, *fractures in limestones helped form Georgian Bay's most famous landmarks – the flowerpots on Flowerpot Island near Tobermory. Such "sea stacks" are produced on cliffed coastlines when broken rock along vertical fractures is eroded. This isolates standing blocks of intact rock that are commonly capped by harder protective rock layers.*

Evidence of a fractured past *The many straight rivers and lake shorelines that surround Georgian Bay follow deep fractures in the crust, a reminder that this part of Ontario was once part of a global supercontinent (Pangea) that, when it broke apart, stretched much of Ontario, fracturing the underlying crust.*

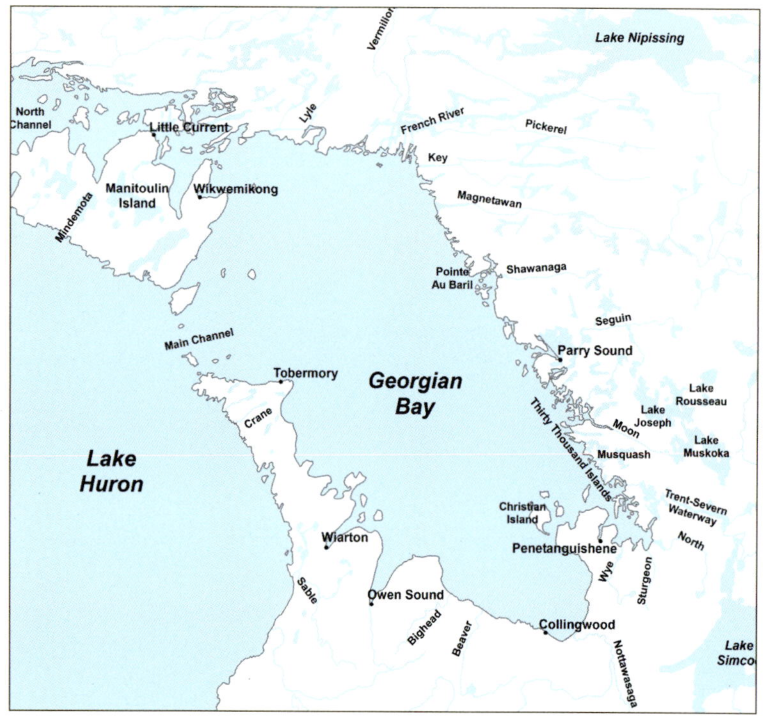

Modern day "tor" near Prescott, Arizona. These landforms would have been common in Georgian Bay before the onset of Ice Ages.

Ontario and US. These boulders are called "erratics," a word derived from the Latin meaning "to wander."

The oil industry comes to Georgian Bay... and leaves

Small "oil seeps" have long been known on Manitoulin Island but the chance discovery of "burning rock" by quarrymen in 1857 at Meaford, on the south shore of Georgian Bay, resulted in recognition that the 400-million-year-old black shales had enough organic content (kerogen derived from ancient algae) to allow distillation and recovery of oil. The recovery process was simple. Rock was heated over fires to 500° C in huge cast-iron distillation retorts. This was a messy, inefficient business that consumed vast quantities of wood only to produce some 250 gallons of oil from 30 tons of oil shale each day. It was also very dangerous and regularly subject to unexpected fires. The discovery of oil in Lambton County, at Petrolia in 1858, made oil shale recovery uneconomic. Not surprisingly, the oil shale industry closed not long thereafter, in 1863. Occasionally, the province casts envious glances at the Meaford oil shale potential and the shale deposits that extend under Blue Mountain but they remain, thankfully, uneconomic. The shale at Meaford is known by geologists as the Collingwood Member of the Lindsay Formation, but instead of oil, the shale here is now better known for its trilobite fossils, first discovered by Alexander Murray in 1845.

Is the Thirty Thousand Islands archipelago unique?

The short answer is yes. There are areas of similar geology and landscape on the so-called Fennoscandian Shield exposed in Norway, Sweden and Finland which has a very similar Precambrian geology of highly metamorphosed gneisses. The Scandinavian rocks are a little older (1.8 billion) and as recently as 350 million years ago, Scandinavia (Baltica) was part of eastern North America before the Atlantic Ocean opened and the two went their separate ways. The

According to Wyandot legend, *the bad-tempered Huron giant Kitchikewana (son of the Great Spirit Manitou), who was overcome by grief and anger at being spurned by the princess Wanakita, hurled a handful of earth into Georgian Bay creating the many islands now known as the Thirty Thousand Islands. Exhausted, he lay down to sleep forming Giant's Tomb Island, the isle of legends.*

so-called "strandflat" region in western Norway consists of thousands of islands and shoals (called skerries) that form a low-relief coastal plain as much as 60 kilometres wide much like the Thirty Thousand Islands area of Georgian Bay.

But there is a major difference: the strandflat is the creation of cold-climate processes and is limited to high sub-polar and polar latitude coastlines where heavy winter sea ice and intense frost activity can gnaw away the shoreline, creating a flat wave-cut bench. High coastal cliffs in Norway mark the inland limit of the strandflat where it has eaten into high plateau dissected by deep glacially-cut valleys (fiords) whose floors now lie more than a kilometre or more under the sea. The strandflat is in essence a young and still evolving landform. Not so the 30,000 Islands which mark where the low relief but bumpy surface of the Canadian Shield simply disappears under the waters of Lake Huron. The Shield surface, though certainly scraped and sand papered by successive ice sheets, contrasts with the strandflat surface by being very old with a long complex history involving warm, wet tropical climates and deep weathering of rock. In many ways it has rebuffed the advances of ice sheets, shrugging them off and remaining largely untouched.

The Finnish Archipelago along the southwestern coast of Finland where the Gulf of Finland opens into the Baltic Sea is perhaps the closest equivalent to the 30,000 Islands. It is a UNESCO World Heritage site and similarly contains some 30,500 islands, the largest of which is Aland. It too, marks where the surface of the Scandinavian Shield (also formed by deep weathering) passes below the Baltic Sea. Strictly speaking therefore it is an example of a marine archipelago unlike the Thirty Thousands Islands of Georgian Bay which is a unique freshwater archipelago.

The many islands of eastern Georgian Bay are the remains of a once giant mountain range more than a billion years old.

CHAPTER

2

THE ICE AGES COME TO GEORGIAN BAY

NICK EYLES

Glacially-striated rock exposed *in front of a modern glacier—what Georgian Bay would have looked like 12,000 years ago as the last great ice sheet retreated northwards.*

Geology doesn't repeat itself but it often rhymes: the Ice Ages

Around 2.5 million years ago the whole world changed. A forested Arctic region with its giant redwoods was no more and Canada became periodically entombed under ice as much as 3 kilometres thick. The modern climate system had begun in which long ice ages (100,000 years) alternate with warm but relatively short (10,000-year) interglacials. We live in the latest interglacial called the Holocene ("wholly recent time") which began about 10,000 years ago when the last ice sheet disappeared from Canada for good. This steady rhythm in climate (paraphrased above by Mark Twain's famous expression) is recorded world-wide.

During periodic ice ages the world's deserts become wet and enjoy abundant rainfalls, while rain forests thin out under cooler, dryer conditions. This was the backdrop to human evolution in East Africa. The oldest hominid (*Homo habilis*, the tool maker) with an enlarged brain first appeared in Africa with the onset of the ice ages. Glacially-lowered sea levels subsequently allowed modern humans to migrate out of Africa moving from one continent to another. We are creatures of the ice ages in more ways than one for as we will learn later in this book, a few hardy descendants of these early emigrants from Africa were to arrive in Georgian Bay via Siberia about 11,500 years ago just as the last ice sheet was leaving.

The earliest geologic evidence for ice sheets in mid-continent North America consists of glacial sediments in Kansas and Missouri, well south of the limits of the last ice sheet. These sediments are rich in clay minerals suggesting the first ice sheets removed deeply weathered rock from the Canadian Shield. The oldest ice sheet grew about 2.4 million years ago and expanded south into the USA leaving sediment called the Atlanta Till but did not expand south again until 1.3 million years ago. Between these dates the ice sheet was no further south than the edge of the Canadian Shield. These ice sheets removed any weathered debris from the Shield leaving freshly-exposed rocks we see today. By about 1 million years ago, the basic ice age pattern of alternating large continental-scale ice sheets and short brief interglacials had been established.

Much remains to be known about the history of the earliest ice sheets in Canada. This lack of knowledge represents a huge gap in our understanding of Canada's recent geological history representing a time when major changes were being made to the Georgian Bay landscape.

The cold war between "Big Ice" and "Big Seas"

It seems strange now but the Great Lakes – including Georgian Bay – were the topic of one of the fiercest debates of nineteenth-century science; that is, the role glaciations played in the formation of the lake basins and landscapes. It was not until 1937 that a glacial origin for the Great Lakes and Georgian Bay was finally accepted. This was exactly one hundred years after the Swiss-born Louis Agassiz argued that continent-scale ice sheets had existed on Earth and altered its

The former existence of ice sheets *in the Northern Hemisphere (Ice Ages) was controversial when first proposed in the 1830s. Deep scratches (called striations) are direct evidence of the flow of ice across the Georgian Bay region during the last ice age.*

Large-scale glacially-cut grooves and ridges on the Canadian Shield near the entrance to the Key River, eastern Georgian Bay.

Small-scale ridges and grooves (striations) made on limestone by rocky debris held in the base of ice flowing from bottom left to top right. Small, more resistant fossils called 'broyozoa' in the limestone and other fossil debris stand proud of the surface as streamlined ridges.

Glacially-transported bolders *("erratics") from northern Ontario, at rest on Brebeuf Island in southern Georgian Bay.*

Right: ***A very distinctive rock type*** *found as erratic boulders is "puddingstone" where red jasper pebbles float in white quartzite. The parent rock is the Lorrain formation (the same as the quartzite of the La Cloche Range at Killarney) that outcrops just east of Sault Ste. Marie. These erratics are found as far south as Ohio and are common along the North Channel at Manitoulin Island.*

landscape. This revolutionary idea was far ahead of its time. English geologist Charles Lyell's *Travels in North America* (1845) utterly rejected Agassiz's notions of large, erosive ice sheets and its dangerous whiff of catastrophism then so prevalent in the revolution-prone continent. He believed that erosion and deposition by marine currents and drifting icebergs (the so-called "drift hypothesis") were the agents of change. His strongest supporter was J.W. Spencer, the leading Canadian adherent of the "glacial-skeptic" school that argued the Great Lakes were the result of vertical movement of the Earth's crust. John William Dawson was another sceptic, declaring in 1878:

> *"with reference to these far travelled boulders, the theory of floating ice serves to account for it as well as land ice and in my judgement, greatly better … The manner in which the great Canadian lakes have been cut out of the softer materials, and their relation to the neighbouring*

CANADA
1989
DOLLAR

harder portions of the country, prove that the great basins may be due to the action of ocean-currents."

Slowly, evidence for an enormous Canada-wide ice sheet was marshalled by field geologists such as Joseph Tyrrell and Robert Bell. Nevertheless the Cold War dragged on until Francis P. Shepherd set the record straight, concluding that the great depth of Lake Superior (some 700 metres below sea level) was the erosional work of powerful ice sheets, not rivers or ocean currents.

The Great Lake basins contain some 23,000 km^3 of water which gives us an idea about the amount of rock that was removed by glacial erosion. Most of this broken and ground-up debris was exported by ice sheets south into the US where it was left as glacial sediments.

Erratic boulders

Large, glacially transported erratic boulders are common across the Georgian Bay basin where they often form lonely sentinels or lurk below the waterline to catch unsuspecting boaters. John Bigsby (1792-1881) was a British Army surgeon who was the first to consider the origin of Lake Huron. He was intrigued by the many large boulders that lay around and were clearly derived from some distances away (then called "rolling rocks"). Along with Henry Bayfield, Bigsby was a confirmed "Neptunist,"that is, a member of a school of geology that believed that rocks were the product of floods (the "Vulcanists" on the other hand, thought they were entirely volcanic in origin). Bigsby proposed that there had been a "great convulsion" of water that flowed out of Hudson Bay, eroding the Huron basin and scattering large boulders everywhere. Glacial deposits, including erratics, were called "drift" because it was believed that such floods carried icebergs. Much later geologists realized that the boulders had been carried by giant ice sheets during the many ice ages. The "drift" concept has long since been abandoned, but the name still survives as a catchall term for glacial deposits.

The Deep Sea Record of Climate: Astronomical Controls on Ice Ages

The discovery of ancient ice ages in 1837 is one of the scientific triumphs of the Victorian age, but it was James Croll, in his famous book *Climate and Time* (1875), who was first to argue that ice ages were controlled by changes in Earth's orbit. This theory was widely rejected because Croll estimated that the last ice age had ended 80,000 years ago. The problem was, data from the Niagara Gorge in Ontario (wherein the end of the last ice age was estimated from the recession rate of Niagara Falls) indicated a much more recent origin for the gorge – some 12,000 years.

We live in an interglacial called the Holocene that finally got underway about 10,000 years ago. The preceding ice age is called the "Wisconsin" while the interglacial before that is called the "Sangamon". Very unusually, Sangamon sediments at least 100,000 years old are preserved in Toronto in the Don Valley Brickyard. This was a time when giant beavers roamed the forests — their strong durable teeth (but not their bones) also frequently turn up in the sediments exposed in the former brickyard.

It is now apparent that there have been at least 50 lengthy ice ages in the last 2.5 million years or so. The land record is sparse because each ice sheet destroys the sediment record of its predecessor. However, we find the full record preserved as successive sediment layers laid down in the deep sea.

The basic climate rhythm of ice ages and interglacials is the product of "astronomical variables". This was finally established by astronomer Milutin Milankovitch in the early twentieth century and confirmed Croll's basic ideas. We are all familiar with an astronomical control on brief seasons and the swings from summer to winter. The ice age cycle is yet another such form of astronomical control on long-term climate. The Earth moves around the sun along an elliptical pathway ("stretch"); its tilted axis around which it rotates alternately increases and decreases in angle ("roll"). The planet is also a bit unsteady as it rotates ("wobble") which is another factor. These three variables – stretch, roll, and wobble—have distinct periods that range from 100,000 years to 23,000 and, when combined, influence how much solar radiation reaches the Earth's surface. The existence of these variables as a major control on Earth's climates has been clearly demonstrated from detailed studies of long sediment cores taken from the world's ocean floors.

The key to ice ages is not extreme cold but humid winters (to allow snowfalls) and cool summers that allow winter snows to survive; the growth of huge snow fields in turn acts like a giant mirror and reflects solar radiation back whence it came. The Laurentide Ice Sheet is the product of the slow but steady expansion and thickening of snow fields over tens of thousands of years.

Huge and ancient boulders lying
Where dissolving glaciers left them.
–Percy Robinson, *The Georgian Bay* (1966)

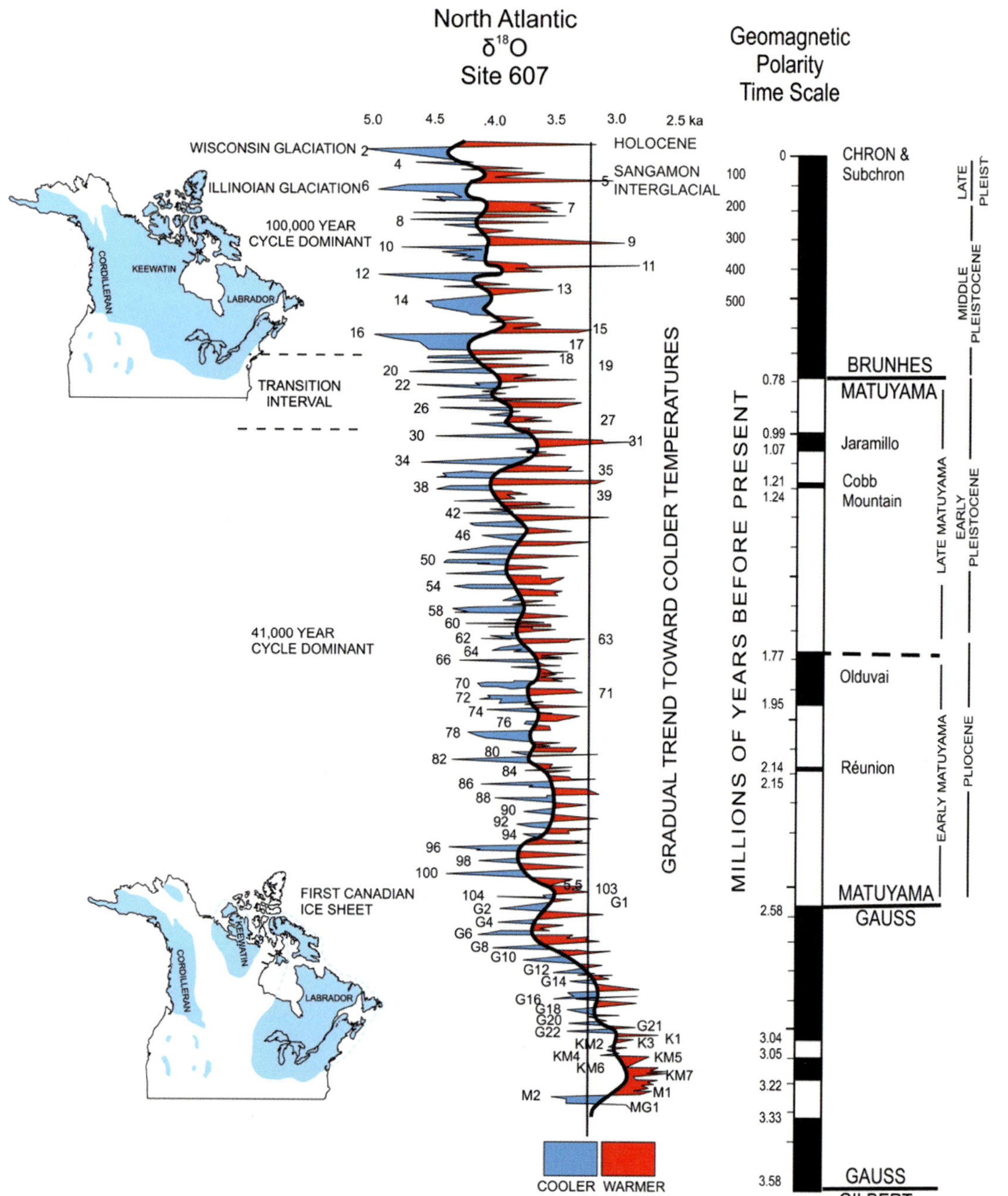

Swings in global climate *from ice ages (in blue and given even numbers) to warmer interglacials (red with odd numbers) over the last 3 million years or so are revealed by studies of sediments accumulating in the deep oceans. Ice ages have gotten longer and cooler through time. The last glaciation to have affected the Georgian Bay region is known as the Wisconsin, and the previous interglacial is called the Sangamon. There is ongoing debate as to when our own interglacial (the Holocene) will end.*

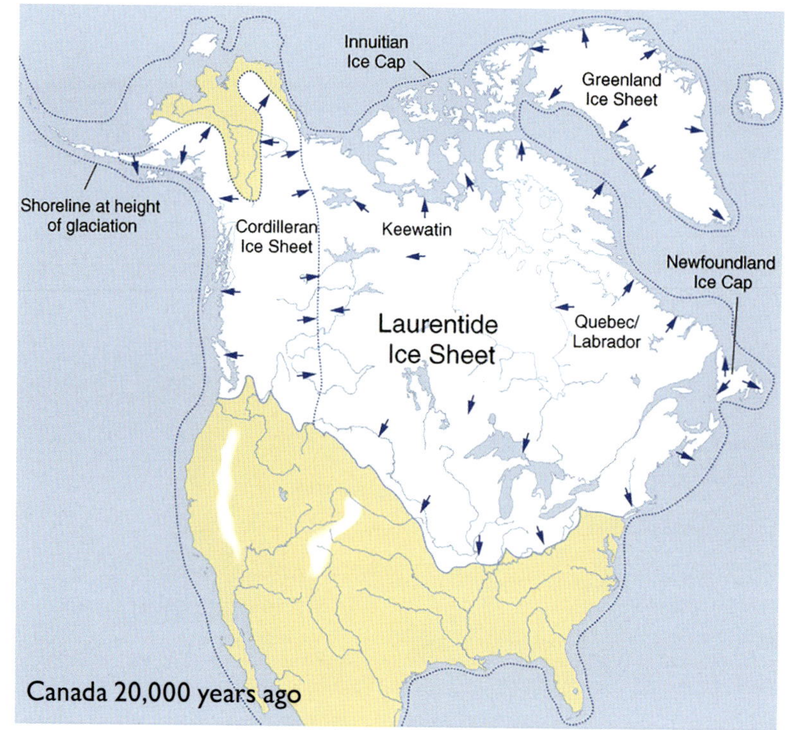

The last great Canadian Ice Sheet

The Laurentide Ice Sheet contained an astonishing 30 million km^3 of ice. It held so much water (that would have otherwise flowed to the oceans as rivers) that global sea level dropped by as much as 70 metres! Its effects were felt as far away as Australia. This ice sheet required tens of thousands of years of winter snowfalls and cool summers to reach its full dimensions about 20,000 years ago. At this point global sea level was at its lowest (it dropped a total of 150 metres because of the other massive ice sheets including those over Europe and Antarctica). Later, when the Laurentide Ice Sheet began to melt very quickly about 11,000 years ago, enormous volumes of freshwater were suddenly released back to the oceans. This had major effects on climates around the world. In mid-continent, some of these meltwaters were briefly trapped in the Great Lake basins to form ice-dammed lakes very much larger and deeper than the modern lakes such as glacial Lake Algonquin whose history we trace in the next chapter.

The role of fast-flowing ice streams

An ice sheet is a deforming mass of ice that flows from its inner areas

to its margins and in so doing erodes its substrate. Present-day ice sheets in Greenland and Antarctica contain corridors up to 100 km wide of ice that flows much faster than surrounding ice. These are called "ice streams." New satellite imagery of Georgian Bay reveals the erosive power of ancient ice streams within the Laurentide Ice Sheet. This evidence comes in the form of giant scratches called *megagrooves* that are several metres wide and up to 1 kilometre long. These are found on the bare limestone plains along the Bruce Peninsula and Niagara Escarpment and from the air closely resemble corrugated cardboard. Megagrooves were also cut into crystalline Shield rocks

A modern glacial lake in Iceland. *This is what Georgian Bay looked like 12,000 years ago when an immense lake called glacial Lake Algonquin was dammed by the retreating margin of the Laurentide Ice Sheet.*

***Aboriginal legends tell us that "potholes"** such as these are the footprints of giants. These intriguing landforms are very common around Georgian Bay and occur on all rock types. They are all relics of the last ice age when huge volumes of turbulent, fast-flowing waters flowed under and away from the melting ice sheet. Potholes were literally drilled into rock surfaces by boulders that were spun round and around in turbulent eddies. The one above is 6 m in diameter and is cut into the cliffed margins of the Key River.*

along the eastern shore of Georgian Bay between the French River and Parry Sound.

Much of the Niagara Escarpment has been sculpted by fast flowing ice into large promontories called "rock drumlins". Their distinctive bullet-shaped profile is most spectacularly seen on Manitoulin Island and along the eastern coast of the Bruce Peninsula and create a zig-zag form to the escarpment when seen from the air. Their presence tells us that ancient Canadian ice sheets were very similar to the modern-day counterparts in Greenland and Antarctica.

Around the southern margins of Georgian Bay fast-flowing ice moulded and sculpted sediment into long ridges called drumlins. A drumlin resembles an overturned row boat with the bow pointing upglacier. Instead of rock, these are cored by till (a Scottish word meaning "stony ground"). The deep soils formed in the tills in the last 10,000 years explain the major contrast between the far south of Georgian Bay, with its prosperous farming communities, and the expanses of bare rock to the north.

The future is cool too

It is a sobering thought that over the course of the last 2.5 million years,

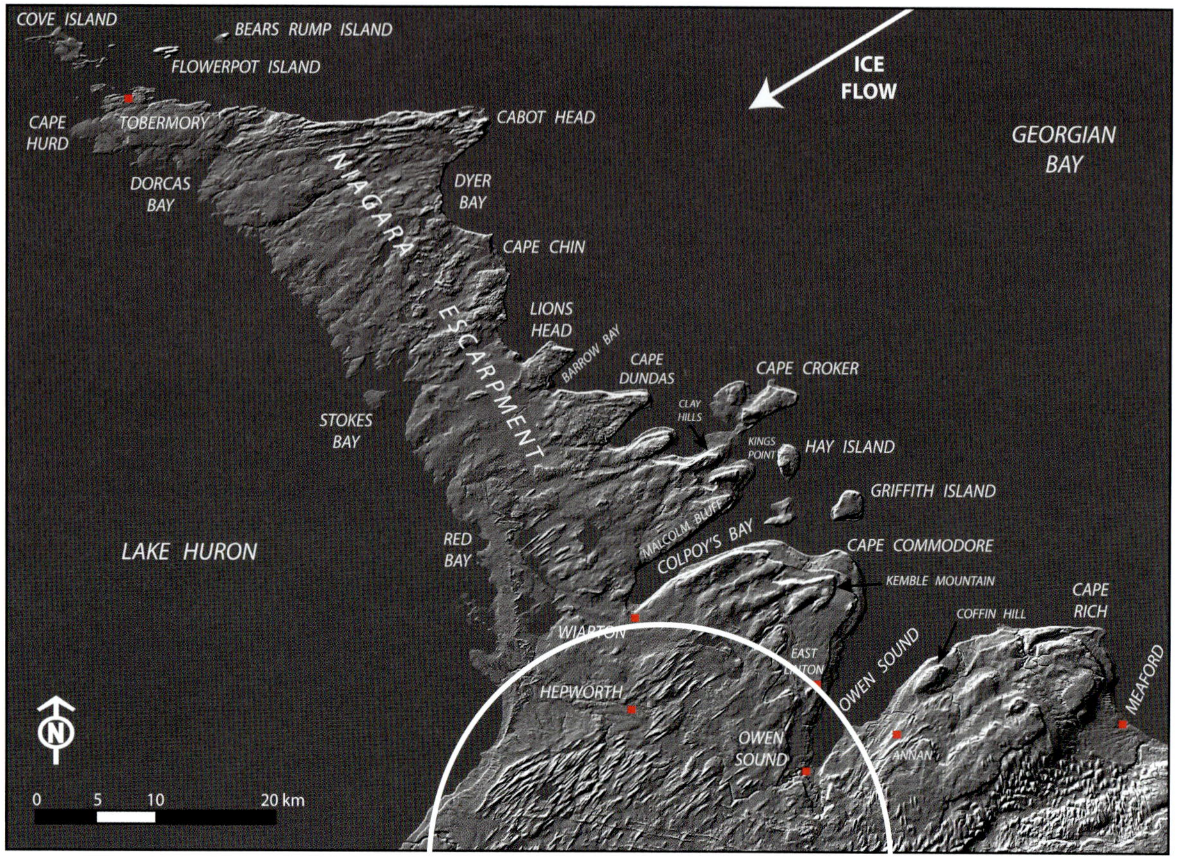

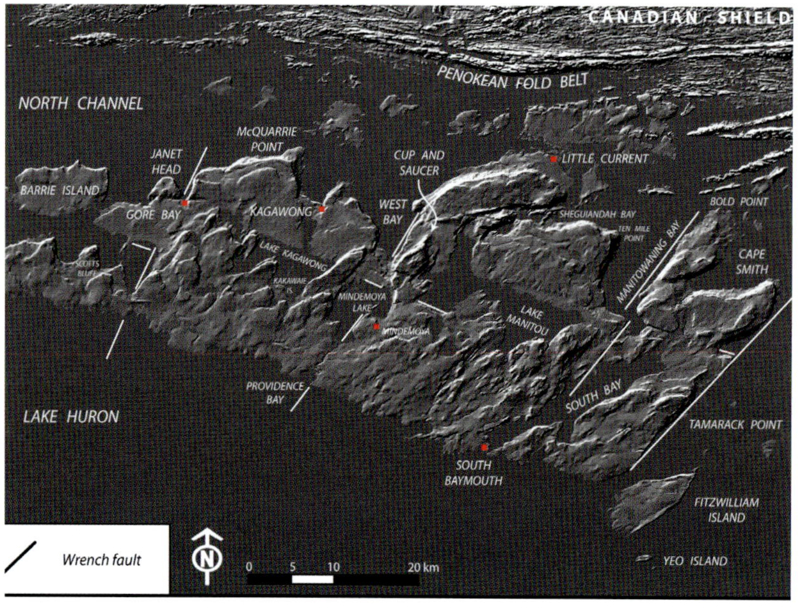

Streamlined hills called drumlins *occur in groups called swarms that contain many hundreds as seen here south of Wiarton (circled), at the base of the Bruce Peninsula. Drumlins record the flow direction of the last ice sheet. The same streamlined landforms carved into limestone of the Niagara Escarpment give rise to the large 'bullet shaped' hills of the Bruce Peninsula and Manitoulin Island.*

the place we call Canada has been buried under thick ice for more time than it has been ice free; our country is truly the home of ice. In fact, Canadians have more to fear from global cooling than warming since we are close to the end of our present interglacial and while human activity and associated global warming may delay the next ice age climate scientists suggest that it may not change the basic oscillation of climate. Georgian Bay awaits the inevitable embrace of the next ice age when it will buried under ice that will sweep it clear of any vestige of human activity. But don't worry, that's thousands of years hence.

A catastrophic flood in Georgian Bay?

Glacial striations, long deep megagrooves, streamlined rock hills (rock drumlins) and many thousands of knobby roche moutonnées are the classic hallmarks of ancient ice ages in Georgian Bay when kilometre-thick ice sheets flowed southwestward across the 30,000 Islands and Bruce Peninsula scraping the rock below. There are also numerous potholes and meandering sinuous grooves cut deep into bedrock by meltwaters flowing in channels both under and in front of the ice sheet. Some geologists have argued that these meltwater-cut features record a gigantic, indeed catastrophic flood of water some 30 m deep and 100 km wide flowing under the last ice sheet; the same 'megaflood' event is even used to explain the many thousands of elongate streamlined hills (drumlins) occurring elsewhere in Southern Ontario, most famously around Wiarton, Peterborough and Guelph. This belief is a throwback to the early nineteenth century when geologists were unaware of past ice ages and resorted to the biblical Noah's flood to refer as 'diluvium' (meaning literally 'of a flood') to what we now know are glacial landforms and sediments.

The notion of an ancient catastrophic megaflood in Southern Ontario has not gained traction among geologists familiar with modern glaciers and ice sheets. They emphasize instead, the importance of subglacial rivers flowing at very high pressures under the last ice sheet, just as is seen today at modern glaciers. Glacial meltwaters are 'dirty' in that they carry large amounts of debris and are very effective in sandblasting bedrock to form potholes and sinuous grooves which are common on the beds of modern glaciers.

Large glacially-streamlined rock knobs *(called 'whalebacks' or 'roche moutonnées') occur in the thousands across the 30,000 islands. Their streamlined rounded, often bullet-shaped noses face up ice: their other ends are rough where ice plucked boulders.*

Chatter marks (at left) are crescent-shaped gouges recording the dragging of boulders over rock at the base of the last ice sheet. Deep glacial megagrooves (top) are also common on the old hard rocks of eastern Georgian Bay. These are the characteristic hallmarks of abrasion below fast-flowing ice during successive ice ages.

This field of parallel streamlined bedrock ridges records abrasion below fast flowing ice during the last ice age.

***This field of glacially-cut bedrock ridges** shows numerous sinuous channels cut by meltwater flowing below the last ice sheet. Surfaces such as this are thought to occur below modern ice sheets and those around Georgian Bay are excellent labratories for studying modern counterparts buried today under thick ice.*

This is a 'pod' of streamlined rock drumlins with rounded noses pointing up glacier. The effects of powerful meltwater rivers flowing under the last ice sheet is recorded by more sinuous channels .

CHAPTER

3

THE UNIQUE FRESHWATER ECOSYSTEMS OF GEORGIAN BAY

KARL SCHIEFER

***The western coast of Georgian Bay** such as seen here near Cabot Head on the Bruce Peninsula consists of high limestone cliffs formed by the Niagara Escarpment. Lake waters here are very clear and cold, increasing in depth rapidly from the shoreline. The result is a very narrow littoral zone and no wetlands.*

The ecosystem of Georgian Bay has many unique features which set it apart from its sister Great Lakes and most of the other large lakes in Canada. This reflects the diverse geological and topographical features surrounding this water body, with high limestone cliffs on the west, extensive sand beaches to the south, and a vast archipelago of bare Shield rock islands of the Thirty Thousand Islands area to the east and north. Its offshore waters are also unique, with an extremely wide littoral zone on the east and increasing water depths toward the Niagara Escarpment to the west.

The water chemistry of Georgian Bay is strongly influenced by the geology of the basin and the tributary watersheds which drain into the lake. Waters draining the Shield on the east tend to be very soft (low in dissolved minerals), acidic, and brown-coloured. By comparison, waters draining limestone to the south and west tend to be harder (more dissolved minerals), basic (non-acidic), and clear (non-coloured). This contrast of physical and chemical features

creates an amazing diversity of habitats around and within Georgian Bay, both aquatic and terrestrial. One of the more important habitat components are the abundant, diverse, and ecologically productive wetlands on the eastern and northern shores. Georgian Bay is also considered to have one of the highest diversities of reptiles and amphibians in Canada, including a number of rare and threatened species such as the Massasauga rattler. This high biodiversity also applies to other plant and animal groups, including birds, fish, and mammals, all of which we can see reflected in conservation efforts to protect them.

No other animal species is so emblematic of the sensitive ecosystems around Georgian Bay as the Massasauga rattler (*Sistrurus catenatus*), a venomous pit viper which occurs primarily along the eastern side of Georgian Bay, on the North Shore of Lake Huron and on the Bruce Peninsula. It occurs as distinct demographically and genetically isolated sub-populations; those at Killbear Provincial

The headland at Lions Head *shows the full height of the Niagara Escarpment. This steep cliff extends into the water, resulting in very deep nearshore waters and a very narrow littoral zone with little to no aquatic vegetation.*

Extreme exposure to *storms and west winds of the offshore islands of Georgian Bay limits plant growth and the development of soils, resulting in a low biodiversity of plants and animals.*

The eastern coast of Georgian Bay *consists of a vast archipelago of islands formed where the Canadian Shield slopes gently into the waters of the Great Lakes and creates a very high diversity of both aquatic and terrestrial habitats.*

Park are distinct from those at Beausoleil Island 50 km to the south. It can live until 18 years old and hibernates below the frost line in rock piles, burrows and tree roots, emerging in late April. It is very dependent on habitats along rocky lakeshores or islands where bogs lie adjacent to upland areas, where it can sun itself (to digest food) and seek protection; all areas favoured by humans and easily disturbed. They are good swimmers. Many are killed on roads where it enjoys the warmth of tarmac. It hunts at night and lives off frogs and small rodents such as deer mice and voles which it finds by heat seeking. It is very docile and endangered because of habitat loss and because of misplaced anxieties about snakes and intentional killing. It is a bellwether for the future of pristine ecosystems of Georgian Bay; as the rattler goes in the next 50 or 100 years, so too does the integrity of the Bay.

The eastern shore of Georgian Bay contains the world's largest and most diverse freshwater archipelago. This area was designated by UNESCO in 2004 as the Georgian Bay Biosphere Reserve. This eastern coast also contains a number of national and provincial parks.

The western shore of the lake is bordered by the limestone cliffs of the Niagara Escarpment which has also been recognized by UNESCO as the Niagara Escarpment Biosphere Reserve. As well, it contains Bruce Peninsula National Park and Canada's first National Marine Conservation Area on the northern tip of the Bruce Peninsula at Fathom Five National Marine Park. The northern shore of Georgian Bay includes Manitoulin Island, which is the world's largest freshwater island and an extension of the limestone cliffs of the Niagara Escarpment and Bruce Peninsula. The southern shore contains an extensive sand beach and dune complex at Wasaga Beach, often referred to as the world's longest freshwater sand beach.

The Physical Environment

The physical components of an ecosystem include its geology, topography, landforms, watersheds, and soils which create the habitats on which the biological components depend. When looking at an aquatic ecosystem involving a lake basin, additional physical components include lake bathymetry (water depths), littoral zones (near-shore shallow water areas), islands, and hydrology (water movements and circulation).The bedrock geology and various glacial deposits surrounding and within a lake basin also have a profound influence on the chemical components of the environment.

In terrestrial systems this includes the type and biological productivity of soils which develop in various areas. In aquatic systems, bedrock geology controls the chemistry of river and lake waters, which in turn have a strong influence on which plant and animal species live in these waters. Water chemistry features also have a strong influence on the productivity of any aquatic ecosystem.

Geology and Topography

The geology of Georgian Bay has already been described in earlier chapters where it has been shown that the lake basin occurs at the

interface between major geological regions. This geological setting is unique to Georgian Bay and accounts for its ecological diversity compared to other Great Lakes.

West Coast

The west coast of Georgian Bay is composed of sedimentary limestones which form the plains of the Bruce Peninsula and extend north through a series of islands to Manitoulin Island, which is the northwest shore of the lake. The Niagara Escarpment is located along the eastern side of the Bruce Peninsula as well as Manitoulin Island, reflected in the majestic limestone bluffs and cliffs along the western shores of Georgian Bay.

East and North Coast

The eastern and northern coasts of Georgian Bay form the largest freshwater archipelago in the world. This is where the Canadian Shield gently slopes to the west and becomes submerged below the waters of Georgian Bay. This large archipelago of islands and shoals provides important and diverse habitats of a very large littoral zone of shallow, protected waters which warm up earlier in the spring and are considerably warmer than the open lake throughout the summer. The storm and wave protection provided by the outer islands and shoals permits growth of a variety of species of aquatic plants in the shallower, warmer waters between the islands and in the numerous sheltered bays. Most importantly, a large number of very high quality coastal wetlands have developed in shallow, sheltered areas throughout the archipelago. These range in size from less than a hectare to hundreds of hectares and provide critical habitat to numerous species.

The many islands provide a range of habitats from barren offshore islands battered by storms and waves to larger and protected inshore islands that support a diverse assemblage of plants and animals.

South Coast

The southernmost coast of Georgian Bay is very different from the other regions discussed above. Here the topography and surficial geology are dominated not by hard rock but by deep deposits of glacial sediments, primarily sand and gravel outwash left by glacial meltwaters that drained the Laurentide Ice Sheet at the end of the last ice age. The most distinctive features are the sand dunes and beaches reworked along the shore and offshore areas of glacial Lake Algonquin sometime after 11,000 years ago. This includes Wasaga Beach, one of the world's longest freshwater beaches.

Natural habitats tend to be more limited along this coast, especially aquatic habitats. The adjacent forests contain a good diversity of Southern Ontario hardwood species, with pines and oaks dominating on sandier sites. In contrast, aquatic habitats are much less diverse than those found along the eastern coast. In many areas the lake shore and littoral zone are dominated by sand bars and flats which are constantly reworked by their exposure to large waves. These shores provide limited habitat for aquatic plants and animals, including most warm water fish species. Other areas have cobble rock

Left: ***The northwestern coast of Georgian Bay*** *abuts against the La Cloche Mountains, with their white quartzite ridges and cliffs. Lakes perched on these ridges are highly acidic, with the result that many are fishless.*

Top: ***The several large islands north of the Penetanguishene Peninsula*** *are the product of glacial deposition. The extensive sand dunes and sand beaches found here are very different from the granite islands of the eastern coast. This is the eastern side of Giants Tomb Island, similar to the landscapes and habitats on nearby Christian, Beckwith, and Hope Islands.*

The lower French River *where it discharges to Georgian Bay, is typical of the larger eastern coast rivers which drain large numbers of lakes on the Canadian Shield. These rivers provide critical spring spawning habitat for walleye and sturgeon in Georgian Bay.*

and boulder shores due to wave exposure, and a littoral zone of rock and sand with little aquatic plant growth. Because of the exposed nature of the coast, there is much less coastal wetland habitat found here. As a result, reptiles and amphibians are also less abundant than is the case along the eastern archipelago.

Lake Bathymetry

Most lakes are shallower around their margins and deepen offshore but Georgian Bay is an exception because of its underlying geology. The eastern and northern coasts are shallow as the resistant granitic bedrock of the Canadian Shield gently slopes westward and gradually sinks into the waters of Georgian Bay forming the Thirty Thousand Islands area. Water depths gradually increase from east to west such that the deepest waters of Georgian Bay (92 metres or 302 feet) occur on the western coast at the base of the steep cliffs of the Niagara Escarpment, along the north-eastern shores of the Bruce Peninsula.

An interesting exception is Parry Sound which was formed by glacial erosion. This bay on the shallow eastern coast of Georgian Bay has a basin (Big Sound) that is 103 metres (340 feet) deep, which is deeper than any of the waters within Georgian Bay itself. This has

become an important cold water ecosystem within the Big Sound, noted for its exceptional lake trout population.

Tributary Watersheds

As might be expected, the diverse topographic and geological features of Georgian Bay's coastline have a strong influence on the tributary watersheds and rivers of each coast and thus their ecology.

On the west coast, there are few tributaries that discharge to Georgian Bay and those that do are very small. The Niagara Escarpment forms much of the coastline here and interior watersheds drain to the west to Lake Huron. Consequently no rivers of any size drain to Georgian Bay.

The east coast by comparison has a large drainage catchment area extending as far east as the highlands of Algonquin Park and Haliburton. Rivers are much larger and include (from south to north) the Severn, Musquash (Muskoka), Moon, Seguin, Magnetawan, Pickerel, and the French.

Because they drain areas on the Canadian Shield, these watersheds contain a large number of lakes, marshes, and bogs perched on hard crystalline bedrock. These waters tend to have some

***The Go Home River** is typical of many of the smaller rivers which drain into the eastern coast of Georgian Bay, in this case originating in Go Home Lake. Even these small rivers provide important spawning habitat for fish.*

light-brown colouration (tannic acids), are acidic to varying degrees, and are low in dissolved mineral content (i.e."soft"). By flowing over bedrock, these rivers tend to be relatively warm in summer while there are few cold groundwater springs during the heat of summer to cool water temperatures.

These rivers provide important spawning and rearing habitat for a number of Georgian Bay fish species such as walleye and lake sturgeon, which migrate upriver in the spring shortly after ice-out to spawn in rocky rapids and on gravel bars. In fall, many of these rivers support spawning runs of Chinook and Pink salmon introduced to the Great Lakes several decades ago.

Tributaries discharging to the southern coast of Georgian Bay are very different than those of the eastern coast. They are much smaller in both watershed area and flow volumes. They include, from west to east, the Sydenham, Bighead, Beaver, Nottawasaga, Sturgeon, and Coldwater rivers. Because they drain an area with deep and extensive glacial sediments, these rivers are sustained by groundwater discharge. This and the presence of underlying limestones control the water chemistry of these rivers. Spring waters tend to be clear, basic (i.e., non-acidic), and contain higher levels of dissolved minerals ("hard" water). Because these rivers are fed by cold groundwater springs, they do not warm up to the same degree as the eastern coast rivers in summer and have some of the best cold-water stream habitats in Ontario. They support resident populations of brook trout and brown trout and have large spawning runs of rainbow trout in spring and Chinook salmon in fall. These river habitats are critical for trout and salmon populations in southern Georgian Bay.

***Rivers on the southern coast of Georgian Bay** are smaller and drain areas of thick glacial sediment. Instead of originating in lakes, like the eastern coast rivers, these streams are maintained by cold groundwater springs. Their gravel and cobble riffles and pools provide important spawning and rearing habitat for Georgian Bay salmon and rainbow trout. This is a cold water tributary of the Nottawasaga River.*

Water circulation in Georgian Bay

As related in Chapter 4, distinct water circulation patterns occur among the many islands of the east coast archipelago of the Thirty Thousand Islands. Because the open waters of Georgian Bay are so extensive, strong winds and changes in atmospheric pressure often move large quantities of water downwind piling up water along the east coast as a "seiche." This produces a corresponding *decrease* in water levels on the western shore. A strong east wind has the opposite effect. Seiches are sometimes referred to as a "freshwater tides" and can result in rapid water level changes of over one metre where water is pushed into the mouths of rivers or narrow inlets. The normal range, however, is between 15 to 25 centimetres.

Channels between the many islands of the archipelago often appear to become rivers as water begins to flow in the direction in which the seiche is moving. This can continue for hours and then reverse itself as water returns back through these channels in the opposite direction. This gives rise to a very complex pattern of currents moving between the islands, depending on the size of the islands, the constriction of the channels, water depths, and other topographic features. This is most variable where large tributaries, such as the French, Magnetawan, and Musquash Rivers discharge into a delta of islands within the archipelago. These currents can be strong enough to prevent ice forming in some island channels for much of the winter making cross-lake travel on ice between the islands a treacherous venture.

In summer, the wind-driven movement of water between islands and in sheltered bays has an important effect on water quality and aquatic habitats. Without these currents and water movements many inshore areas could become too warm and stagnant during summer. These currents disperse nutrients which are important for aquatic plant growth and modify water temperatures in adjacent areas. It is not uncommon to be swimming in relatively warm water one day and significantly cooler water in the same location the next day because of the currents and the mixing of warmer inshore waters with colder offshore waters.

These patterns of water movement and mixing are a unique feature of the aquatic ecosystem of eastern Georgian Bay.

Water Chemistry

There has been a great deal of research and monitoring of water quality in Georgian Bay over the past four or five decades. Much of this is focused on protecting human health related to drinking water sources and water-based recreational activities. The primary interest here is in water chemistry as it relates to natural ecosystem functions and aquatic productivity.

As identified above, bedrock geology and, to a lesser extent, the glacial history of the region have a strong influence on water chemistry in various areas of Georgian Bay. Waters draining the limestone bedrocks of the western shore tend to be very clear, basic (high alkalinity), and high in total dissolved solids or minerals ("hard"

***Water chemistry profile** on the Georgian Bay coast. The slow and restricted mixing of runoff waters and open lake waters results in water chemistry gradients in many bays and inlets along the eastern Georgian Bay coast. Here, waters in this inlet become clearer, less coloured, less acidic (reflected in higher pH numbers) and higher in alkalinity and dissolved minerals as one moves west toward open waters. Water conductivity (CON) provides a measure of total dissolved solids in the water. Here, CON is low in tributary waters ("soft" water) but increases rapidly through the inlet to the natural background level of 185 for Georgian Bay waters. Similarly, pH is quite acidic at 5.3 in tributary waters (due to the acidic bogs that they drain, as seen in the foreground) but increases rapidly to basic levels above 8.0. This is due to the high natural buffering capacity of Georgian Bay waters. Finally, water clarity is low at 1.3 metres near the inflows and increases progressively to over 9 metres beyond the influence of mainland and island runoff and wave scour.*

water). By comparison, waters draining hard crystalline rocks of the Canadian Shield on the east and north shores have a light brown colouration due to tannic and other acids originating from the bogs and wetlands on the Shield. These waters also tend to be more acidic because of the acid bog drainage and because the bedrock provides little natural buffering for acids. Because Shield rocks are also more resistant to weathering than softer limestones, Shield drainage waters tend to have lower dissolved mineral content ("soft" water).

In most large lakes, the water chemistry is relatively uniform throughout the main lake basin because of wind-driven mixing. This is also true for Georgian Bay, but with one significant and interesting exception. Because rivers draining the Canadian Shield on the eastern coast flow into Georgian Bay within the large island archipelago, mixing of waters is much more variable and less complete. This results in water chemistry gradients within most of the larger bays and channels along the eastern coast. A good example is at Cognashene, a bay on the south-eastern coast of Georgian Bay. Typical of many inlets along this coast, it receives runoff waters from numerous acidic bogs within the watershed and has numerous narrow or shallow channels connecting it with the open lake. These tend to restrict the degree of water mixing in various parts of the bay, resulting in a water chemistry gradient throughout the bay. Closer to the mainland runoff sources, runoff waters have a light-brown colouration, are less clear, are acidic and low in dissolved minerals. Closer to the open lake waters, the water chemistry of Georgian Bay prevails: these waters are clear and colourless, less acidic (or basic), and higher in alkalinity and dissolved minerals.

Those inlets which also receive the discharge of larger rivers, such as the French, Magnetawan, Moon or Musquash, will have an even greater area of Shield inflow water chemistry influence, especially in the spring when river flows and general runoff volumes are greatest.

As runoff and river flows decline through the summer, eastward mixing of lake waters becomes more pronounced due to wind and seiche-related water movements.

Human Impacts on Water Quality

Water quality in Georgian Bay is relatively pristine in most areas, with some exceptions such as Severn Sound, in the southeast corner of Georgian Bay, which has been the site of a water quality remediation program over the past two decades. This is one of the few areas of Georgian Bay with significant urban development on its shores (e.g., Midland, Penetanguishene, Victoria Harbour, Port McNicoll and Port Severn), and watersheds affected by agricultural runoff. These sources have increased nutrient levels (eutrophication) in the waters of Severn Sound, which can result in excessive plankton blooms, aquatic plant growth, and reduced dissolved oxygen levels. That said, water quality has improved in Severn Sound over the past 10 years.

The other area of concern involves some of the bays and inlets along the eastern Georgian Bay coast which have higher densities of lakeshore cottage development but reduced levels of water exchange with the open lake. Examples include Sturgeon Bay, near Pointe au Baril, where late summer blooms of blue-green alga have caused serious water quality issues in recent years. Ongoing water quality monitoring programs along this coast are important if future water quality problems are to be avoided, especially as lakeshore development continues.

Thermal Stratification

One of the important features of freshwater lakes is the summer thermal regime with depth. This has a major effect on habitats for many species, especially fish.

As surface waters warm in spring, a thermocline (zone of rapid water temperature change where waters are warmed to temperatures above 4°C) develops in the water column to depths of three to five metres below the water surface. As related in chapter 4, below the thermocline, waters remain cold all summer (< 4°C) as the warmer, less dense waters of the thermocline prevent mixing of warmer surface waters with these deeper, denser cold waters. As the summer progresses, surface waters continue to warm, pushing the thermocline deeper. By late summer, the thermocline can be 25 to 30 metres below the surface of Georgian Bay. As surface waters cool in fall, the thermocline breaks down and colder lake waters can again mix from top to bottom.

On the western coast of Georgian Bay, thermal stratification occurs very close to the coast because of the very deep waters close to shore. With the thermocline directly abutting this steep coast, the warm water littoral zone (shallow nearshore waters) tends to be very narrow and small in area. As a result, warm-water fish such as

***The islands in the archipelago** near the mouth of the French River provide a diverse range of habitat conditions – from exposed bare bedrock outer islands, to somewhat larger and more sheltered middle islands with sparse vegetation, to even more sheltered interior islands and the mainland coast where more dense and diverse vegetation exists. Aquatic habitats also show this gradient from exposed bedrock shoals with no aquatic plant growth (foreground) to sheltered interior bays with extensive aquatic plant growth and coastal wetlands.*

Offshore islands are sparsely vegetated and the shallow *littoral zone between the islands is also mostly bare bedrock. These extreme aquatic habitats are exposed to storm surges, wave scouring, and spring ice piling which prevents the growth of aquatic plants. As a result, this habitat is relatively low in biological productivity, including fish.*

Left: ***Because of their harsh environment,*** *offshore islands have a very low diversity of plant species. Orange and grey lichens are able to grow on the exposed granite while a few shrubs and hardy perennial plants can grow in some of the deeper fractures.*

Some of the larger offshore islands *support shrubs, grasses, and hardy perennial plants in deeper cracks and bedrock depressions where organic material has slowly accumulated over many decades. These depressions are then able to store rainwater for the inevitable summer drought conditions which affect these exposed bedrock islands. This photo is taken within the Pine Islands, a number of which are protected by the Georgian Bay Land Trust.*

Life is hard on the outer islands. *This stunted pine tree is likely 30 to 40 years old and has survived winter storms and summer drought living in this bedrock crack.*

largemouth bass and northern pike are low in abundance compared to cold-water species such as whitefish and lake trout.

On the eastern and northern coasts of Georgian Bay, however, extensive areas of shallow water between islands and in bays and inlets limits thermal stratification. Here a very wide littoral zone with warm waters from surface to bottom develops during the summer months. The shelter provided by the islands also allows for the growth of aquatic plants in many areas, including coastal wetlands.

***Vegetation** on many islands in Georgian Bay is limited by two environmental factors, lack of soil and summer drought. Pine and other trees growing on insecure sites tend to have die-offs every 5 to 10 years when more extreme summer droughts deplete the limited moisture supply in shallower bedrock depressions. Those growing in deeper depressions often outgrow the site when the root system can no longer support the mass of tree trunk and branches, and the tree blows over in a windstorm. Drought-killed trees and wind-throws with root systems peeled off the bedrock are common features on these islands.*

This provides excellent and extensive habitat for a wide range of warm-water fish species, including largemouth and smallmouth bass, northern pike, muskellunge, walleye, yellow perch, and many others. Cold-water fish species are found farther offshore along this coast, dependent on deeper, cold-water habitats.

An important exception is the Big Sound of Parry Sound and adjacent waters. These are some of the deepest waters in all Georgian Bay and are thermally stratified all summer. As a result, they provide excellent cold-water fish habitat which supports some of the highest quality native lake trout and whitefish populations.

Ecosystem Diversity

Humans have become a major factor in modifying and altering the natural systems within the Georgian Bay ecosystem. Anthropogenic changes in forest cover, river hydrology, water quality, exploitation, and extermination of many native species and the introduction of

***Winter conditions** on the outer and middle islands are extremely harsh. Many of these islands are totally drifted over by winter snow storms, providing no overwintering habitat for reptiles, amphibians, birds, and mammals. Spring breakup also pushes ice over many of the lower islands. This is Lizard Island in late February.*

On some of the larger islands *and the mainland coast of eastern Georgian Bay there is sufficient soil and moisture to support much more diverse vegetation, including numerous species of coniferous and hardwood deciduous trees. In turn, this supports a much higher diversity of animals, including mammals, birds, reptiles, and amphibians. This is near the coast of Massasauga Provincial Park. Similar habitats are found on Sandy Island, Beausoleil Island, and other large coastal islands.*

Unlike the offshore and middle islands, *the larger interior islands and mainland coast of Georgian Bay provide suitable overwintering habitat for many species of mammals, birds, and reptiles. Hibernating species, such as black bear, grey squirrels, and Massasauga rattlesnakes also find suitable habitat on these larger interior islands.*

This is an example of a coastal marsh *occupying the shallow and sheltered water of the eastern Georgian Bay coast near Parry Sound. These marshes provide very important spawning and juvenile rearing habitat for warm-water fish species such as largemouth bass, northern pike, and muskellunge. Waterfowl, shorebirds, aquatic mammals, and many species of reptiles and amphibians depend on such habitats.*

large numbers of non-native, invasive species have greatly altered these natural systems. Sadly, this process is ongoing.

The term "biodiversity" refers to the diversity of numbers and types of plant and animal species which occupy an ecosystem. Because biological communities are directly dependent on their habitats, a greater diversity of habitats in any ecosystem generally results in greater biodiversity of plants and animals. Georgian Bay provides a very high level of habitat diversity for both terrestrial and aquatic ecosystems, which results in a correspondingly high diversity of both plant and animal species.

The large archipelago of islands, shoals, bays, inlets, and river deltas along the eastern and northern shores of Georgian Bay provide the best example of this. Islands provide a gradient of both aquatic and terrestrial habitat conditions from almost bare bedrock exposed

Emergent, floating, and submerged plant *communities are abundant and diverse in this high quality wetland habitat.*

Left: ***Many sheltered bays*** *provide conditions for the development of much larger coastal marshes, as seen here in Go Home Bay. These are some of the most biologically productive habitats in the entire ecosystem, supporting an exceptionally high biodiversity of plants and animals. As an example, the eastern Georgian Bay coast boasts the highest concentration of reptile and amphibian species in Canada. This is also excellent moose habitat.*

Low water levels have a negative effect *on many of Georgian Bay's coastal marshes. Although most of these plants eventually recover or recolonize when water levels rise again, many of the habitat functions for fish, aquatic mammals, and some reptile and amphibian species are reduced or lost during such low water periods.*

to harsh storms on the outer islands to diverse, mixed hardwood forests on the larger interior islands and mainland coast. As a result, biodiversity is much greater on the larger interior islands and mainland coast than on the offshore islands.

Wetlands and Coastal Marshes

Wetlands and coastal marshes provide some of the most important and diverse habitat features in any freshwater ecosystem and form a complex transitional interface between terrestrial and aquatic environments where shallow, warm waters receive nutrients from adjacent forest runoff. The result is an area of high biological productivity and diversity of plants and animals.

Georgian Bay possesses the largest number and some of the highest quality natural wetlands of any of the Great Lakes. This claim could likely be extended to include all freshwater lakes in Canada or North America. The abundance of islands, shoals, bays, and inlets found within the world's largest freshwater archipelago creates an abundance of opportunities for wetlands to develop. The

Interior bogs *occupy shallower depressions on the granitic bedrock of the Georgian Bay coast. Originating as shallow lakes after the last glacial retreat, they have been accumulating organic material, such as peat, for centuries. These are important habitats for aquatic mammals, waterfowl, and numerous reptile and amphibian species. A beaver den is seen in the middle of this photo. These are important reservoirs of water which sustain lake levels and river flows on the Shield.*

generally shallow waters and sheltered shorelines throughout this region have resulted in the establishment of many hundreds of coastal wetlands of various sizes and forms.

These wetlands are one of the critical natural features which support the exceptionally high diversity of plants and animals found on eastern Georgian Bay. This includes a large number of riparian (lakeshore) and littoral zone aquatic plant communities and species which occupy the riparian, emergent, floating, and submergent vegetation zones of a wetland. This provides habitat for a large number of mammals, birds, reptiles, amphibians and fish.

Interior Bogs

Interior bogs are abundant along the Georgian Bay coast and on many of the larger islands. They are essentially shallow water-filled

depressions on the granitic bedrock of the Canadian Shield. Over thousands of years they have accumulated a large amount of organic material from the plant growth within and adjacent to the bog, resulting in thick deposits of peat on which many bog plants grow.

Interior bogs contain water that is stained brown and acidic due to tannic acids from plant decomposition. Bog water is also very low in dissolved minerals due to the non-soluble and weathering-resistant nature of the granitic bedrock on which bogs exist. Interior bogs are important reservoirs of water from precipitation and spring snowmelt. The large number of these in many shield watersheds contributes to maintaining and regulating the summer flow in rivers draining large areas of granitic bedrock with little groundwater discharge. Unlike coastal marshes where water level is determined by the larger lake system, most bogs have more stable water levels which are high after

***Some of the larger islands** also contain acidic bogs perched on bedrock, as seen here. The water is tea-coloured and more acidic than Georgian Bay, but these bogs also support a diverse community of emergent, floating, and submerged aquatic plants. Some of these, such as the carnivorous Pitcher Plant, have adapted to grow in acidic bogs.*

Rivers on the southern coast of Georgian Bay *support large spawning runs of rainbow trout in the spring and Chinook salmon in the fall. This photo shows a group of Chinook salmon spawning in the upper Nottawasaga River.*

The very large littoral zone *created by the island archipelago on the eastern coast of Georgian Bay provides a very productive aquatic ecosystem in these sheltered, shallow and warmer waters. Many nearshore fish species utilize these habitats for shelter, reproduction, and feeding.*

spring runoff and decline slowly through the summer months. Many bogs have water levels which are determined by beaver dams.

As with the coastal marshes, interior bogs support a diverse range of aquatic plant species, some of which are specifically adapted to growing in these acidic environments. They also provide important habitat for aquatic mammals, such as beaver, muskrats, mink, otter and moose, waterfowl, and many species of reptiles and amphibians. The close proximity of interior acidic bogs to coastal marshes in many areas along the Georgian Bay coast further enhances the high diversity of aquatic and riparian habitats.

Fish

The diversity and quality of aquatic habitats in Georgian Bay result in equally diverse fish populations. Habitats include deep, cold and clear offshore waters; an expansive littoral zone along the eastern coast with extensive islands, shoals, reefs, and bays; sand reefs and flats along the southern coast; abundant coastal wetlands and zones of aquatic plant growth; and a large number of high quality tributaries in which migratory fish can spawn. In total, over 85 species of fish have been documented in Georgian Bay. Freshwater fish communities consist of those found offshore (cold-water) and those found nearshore (or warm-water). Typically trout, salmon, whitefish, burbot, and other species which prefer colder summer waters occupy the deeper offshore zones of the lake while the pike, bass, walleye, sunfish, and perch species prefer the shallower and warmer nearshore waters of the littoral zone.

Large fish species found in the Great Lakes.

The extensive shallow waters *of the eastern coast provide an abundant food base for numerous nearshore fish species. Crustaceans (includes crayfish), molluscs (clams and snails), insect larva, and other organisms thrive in this productive aquatic habitat.*

Offshore Fish Community

The top of the offshore food chain was occupied by two cold-water predator species, the lake trout and burbot (sometimes called freshwater cod). The walleye can also be an offshore predator but generally in the cooler waters above the thermocline. These species fed on a forage fish base which included lake herring (cisco), whitefish, sculpin, and a group of deep-water cisco species. The lower levels of the offshore food web depend largely on plankton production in the upper water column and small invertebrates and insect larvae occupying the deeper lake sediments.

When Europeans first arrived on the shores of Georgian Bay, the offshore fish community was dominated by a large and stable population of lake trout, including some fish weighing over 30 kilograms. The offshore fish community has changed

Historic offshore (cold water) fish food web in Georgian Bay

Current offshore (cold water) fish food web in Georgian Bay

Historic nearshore (warm water) fish food web in Georgian Bay

Current nearshore (warm water) fish food web in Georgian Bay

significantly over the past century because of human activities. The introduction of the parasitic sea lamprey combined with overfishing had depleted lake trout populations by the middle of the 1900s. The introduction of alewife and smelt coincided with the demise of several deep-water cisco species, resulting in a significant change in the forage-fish species base. Chinook salmon, rainbow trout, and brown trout were introduced to fill the ecological void left by the native lake trout. Cormorant populations also increased dramatically over the past several decades, and are a significant predator of this fish community. To further destabilize this offshore food web, the introduction of quagga and zebra mussels, along with invasive planktonic species, appears to have reduced the overall biological productivity of this fish community. Unfortunately, this ecosystem remains destabilized as the ongoing effects of invasive species and a changing climate continue to modify the ecological role and abundance of each species.

Introduced rainbow trout, brown trout and Chinook salmon have established self-sustaining populations in Georgian Bay, utilizing the high-quality cold-water tributary streams along the southern coast of the lake. The Coldwater, Sturgeon, Nottawasaga, Beaver, Bighead, Sydenham, and other southern rivers support large spawning runs of rainbow trout in the spring and Chinook salmon in fall. These are spring fed rivers with cold, clear water flowing over gravel and cobble substrates used as spawning habitat by these species. Some of the eastern coast rivers also now appear to support fall Chinook salmon runs.

Nearshore Fish Community

Because of the very large and diverse littoral zone on the eastern and northern coasts of Georgian Bay, the nearshore fish community consists of many warm-water fish species at all levels of the food web. The abundance of islands, shoals, reefs, bays, and coastal wetlands provides excellent nearshore fish habitat.

The dominant predators in this community include the northern pike, musky, walleye, largemouth and smallmouth bass, bowfin, and gars. They feed on an equally abundant and diverse prey-fish community which includes suckers, perch, sunfish, catfish, and many species of darters, minnows, shiners, and other warm, shallow-water species. The base of this food web is supported by primary production largely in the form of rooted aquatic plants and free-floating plankton.

__The walleye is an iconic fish__ species of the eastern coast. The availability of excellent spawning habitats in tributaries such as the Severn, Moon, Magnetawan, and French Rivers combined with an extensive and productive littoral feeding zone in the island archipelago once supported large populations of this species.

Secondary production in the lower food web includes aquatic insects, molluscs (clams and snails), crustaceans (crayfish), and other life forms that utilize aquatic plants and the lake bed.

Human activities have also modified the nearshore fish community but not to the same degree as offshore. The top predators all remain but, in some cases, in much lower abundance. Cormorant population increases over the past several decades have also had a predation effect on the nearshore fish community, especially on some of the smaller prey species and juvenile predators. Carp and round goby are two invasive fish species which have established large populations in Georgian Bay and will continue to influence the ecology of the nearshore zone. As well, the zebra and quagga mussel, which were introduced through ballast discharge from ocean-going ships, have had a major effect on aquatic food chains and nutrient dynamics, and will continue to do so.

One species of special note is the walleye, or pickerel as it is often called. Locally, this species was abundant on the Georgian Bay coast and had large spawning runs up the Severn, Musquash, Moon, Magnetawan, Pickerel, French, and numerous others eastern coast tributary rivers. Unfortunately, overfishing, modified river flows, and ecosystem changes have greatly reduced the populations of this iconic Georgian Bay species. Attempts to restore walleye stocks have been underway for several years by the Eastern Georgian Bay Stewardship Council.

Reptiles and Amphibians

The Georgian Bay coast boasts the highest concentration of reptile and amphibian species in Canada. This concentration is directly related to the very high diversity and quality of terrestrial and aquatic habitats on the Georgian Bay coast, especially the east coast archipelago which ranges from dry, hot bedrock barrens to extensive coastal marshes and perched bogs.

There are at least 34 reptile and amphibian species in the Georgian Bay coastal ecosystem, many of which are at or near the northern range of their distribution. This includes a number of reptile species which are rare or considered at risk, such as the five-lined skink (Ontario's only lizard), the Massasauga rattlesnake, eastern foxsnake, eastern hognose snake, milksnake, spotted turtle, northern map turtle, stinkpot, and Blanding's turtle.

Snapping Turtle

The distribution of these species within the island complex is directly related to habitat conditions. The barren bedrock outer islands generally have very few of these species while the larger and more diverse interior islands and mainland coast have an abundant array of reptiles and amphibian species. This is especially true in close proximity to coastal marshes and perched bogs. These areas provide habitat for reproduction, feeding, and critical overwintering hibernacula for some of the snakes.

Most of these species have relatively small home ranges for their entire life cycle, but there are notable exceptions. The Massasauga rattlesnake, for instance, can often be observed swimming from island to island in the summer, though most return to their same winter hibernation habitat. The eastern foxsnake undertakes even greater summer migrations among the islands.

As related above, the Massasauga rattlesnake is the species most commonly identified with and considered emblematic of the rugged Georgian Bay coast. Its current distribution in Ontario and Canada is largely restricted to the coasts of Georgian Bay, especially the eastern and western coasts. It is not a common species and its long-term survival depends on protection of its habitat.

Turtles are common in Georgian Bay and the largest is the snapping turtle (*Chelydra serpentina*) which can reach over 30 kg in weight and as much as 36 cm in length. It is commonly seen around the eastern shores of Georgian Bay in areas of wetlands, bogs or marshy shorelines. Though capable of defending themselves on land they have an ill-deserved reputation as being harmful to swimmers. They eat small fish, frogs, and crayfish, snails and occasionally dead animals. Mature females produce up to 40 ping pong ball-sized eggs in the fall but few young reach adulthood. They can live as long as 75 years hibernating between October and May but few attain this age because of increasing predation of their eggs by raccoons and skunks whose numbers are rising in response to opportunities arising from urban development. Because of this the snapping turtle is listed as a species 'at risk.'

Wye Marsh is one of the largest wetlands in Georgian Bay and lies near the town of Midland through which the Wye River flows before it enters Georgian Bay. As a wildlife reserve, the Marsh harbors many species including the rare trumpeter swan. Close by is the reconstructed 17th century village of Sainte-Marie among the Hurons which was the western terminus for the early fur trade. The original was burned in 1649 by invading Iroquois warriors and was reconstructed in 1964 on the banks of the Wye River close to the Wye Marsh.

The eastern foxsnake *is one of the largest and most versatile snakes along the Georgian Bay coast. It can not only swim long distances but also climb trees to feed on birds. Like the Massasauga rattlesnake, eastern hognose snake and others, it is considered to be a species at risk and requires protection. This foxsnake was over 1.3 metres long.*

CHAPTER

4

THE UPS AND DOWNS OF LAKE HURON AND GEORGIAN BAY

NICK EYLES

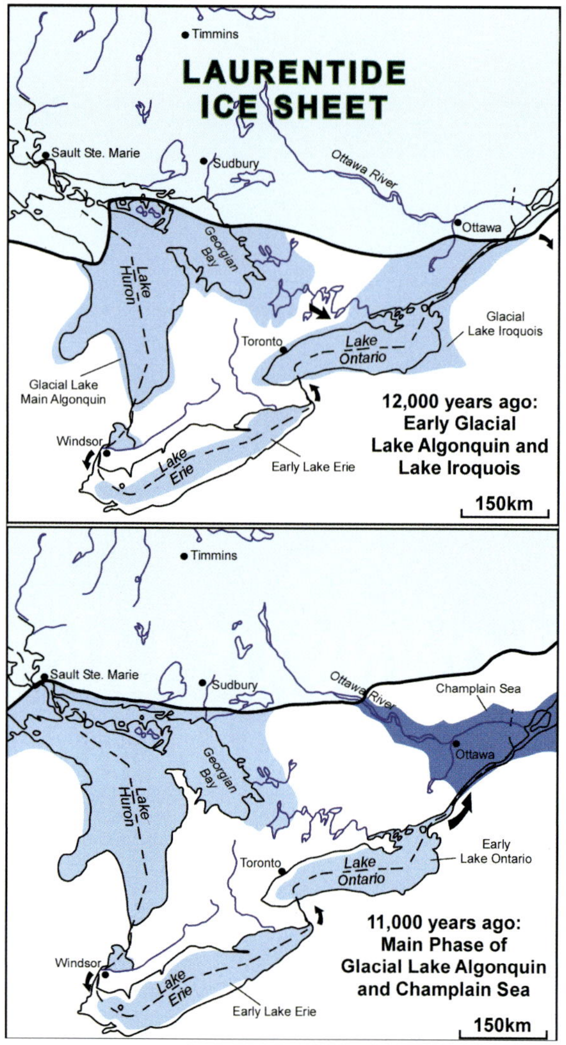

Lakes along the retreating ice sheet margin.

Gifts of the Ice Ages

The lakes we now call Huron and Georgian Bay are the result of the ice ages when powerful glaciers excavated their rock basins which now trap the lakes. Today, relatively small seasonal and decadal ups and downs in lake level are often the source of animated conversations among boaters and cottagers. However, these short-term fluctuations are nothing compared to the massive changes in lake levels over the last thousands of years. Contemporary seasonal changes are no greater than a metre but fluctuations of many tens of metres are recorded in the ancient past.

At the end of the last ice age some 12,000 years ago the retreating ice sheet blocked the outlets of the modern Great Lakes and a large ice-dammed glacial lake (Lake Algonquin) tens of meters deeper than the modern lake formed in the Huron Basin. With further climate warming after 9,000 years ago, lake levels fell below modern levels and much of the floor of Lake Huron and Georgian Bay was exposed to human occupation. This was the time of Lake Stanley in the Huron Basin and Lake Hough in Georgian Bay. Here low but fluctuating lake levels lasted through until about 4,000 years ago when climate cooled and became wetter in central Canada. This filled Lake Huron to its modern size.

Glacial Lake Algonquin (12,000 to 10,400 years ago): the ice sheet retreats from Georgian Bay and humans arrive

During the last ice age, Georgian Bay lay buried under kilometres-thick ice for several tens of thousands of years only to emerge when the northward retreating ice sheet margin freed Georgian Bay about 12,000 years ago. But the outlets of the Great Lakes along the St. Lawrence Valley remained blocked, damming up huge "Greater Lakes." Huge bodies of cold, iceberg-infested meltwaters were dammed up in the Huron (glacial Lake Algonquin) and Ontario (glacial Lake Iroquois) basins and their raised shorelines and beach bluffs are a noticable feature of the modern landscape.

***Lake trout** are the top fish predators in Georgian Bay and are survivors of the last ice age. Its numbers are a good indicator of the health of the wider lacustrine ecosystem. Its name namaycush is a Cree word and it was an important food source for early peoples. It inhabits cold, oxygen-rich waters and spawns inshore on rocky bouldery substrates that allow protection for eggs, but as soon as these waters warm up in early summer, trout move offshore to deep cold waters.*

The earliest humans (Paleo-Indians) who wandered across this landscape 11,000 years ago were met with a grey seemingly inhospitable visage much like parts of modern-day Patagonia or Iceland. Spruce parkland in the south gave way northward to a fringe of tundra willow around the ice margin. This incredible landscape was dominated by enormous ice-dammed lakes, fast flowing and very dangerous meltwater rivers, great expanses of bare rock and newly deposited glacial sediment, and frequent sand storms. Ancient Algonquin beaches, now high and dry, together with swaths of sandy and muddy lake sediments, locally cover the Georgian Bay interior and create rare opportunities for farming on the otherwise rocky expanses of the Shield. In fact, farming of Algonquin sediments was a key factor in the growth of early communities such as Parry Sound, Bracebridge, Huntsville and Gravenhurst.

Humans were not the only species to have invaded the Georgian Bay area as it began to be exposed by the retreat of the last ice sheet just after 12,000 years ago. Lake trout (*Salvelinus namaycush*) are much sought after by recreational fishers and are also famous among ecologists as a glacial relict of the last ice age (along with the deep-water sculpin, *Myoxocephalus thomsoni,* and the opossum shrimp, *Mysis relicta* on which lake trout feed). Their fossil remains are found in the muddy sediments of the large glacial lakes, such as glacial Lake Algonquin. The geographic range of the hardy lake trout extends along thousands of modern-day lakes across mid and northern North America. This area is almost exactly that of the last ice sheet and tells us that the large iceberg-infested lakes that formed along the front of the retreating ice allowed these cold water fish to migrate across all of the Shield. Lake trout first evolved about 3 million years ago and it is no coincidence that this occurred just when the first ice sheets were forming across Canada when global climates cooled. It is a pre-eminent Canadian icon better at home in the cold than in the brief warmth of the interglacials which separate ice ages.

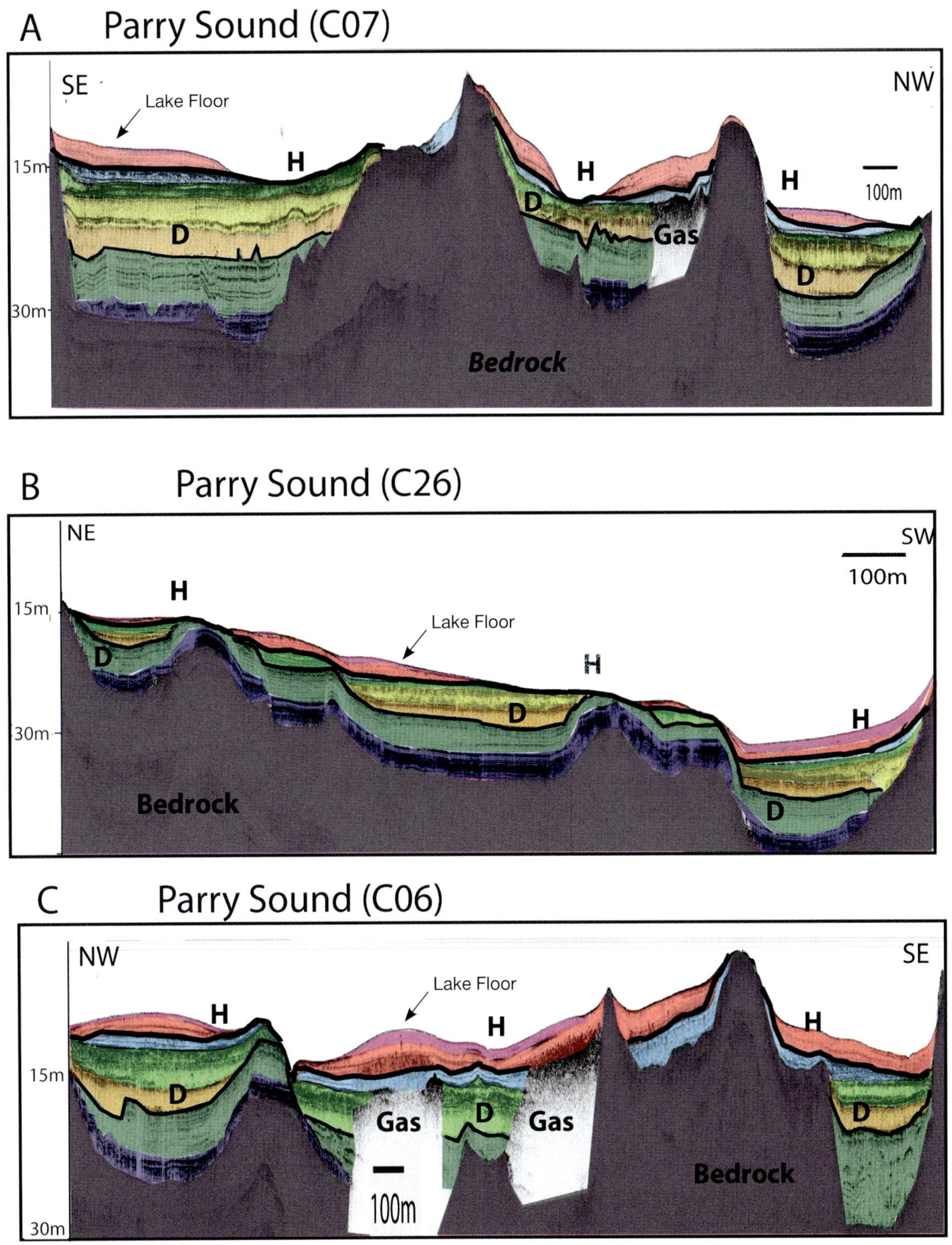

Lakes cover some 21% of the total area of Ontario *(the name means "land of shining waters"). The scientific study of lakes and their dynamics is known as "limnology" and the study of ancient lakes is known as "paleolimnology." The ups and downs of lake levels over thousands of years tell us much of past climates and also provide clues about the future. Lake basins are natural "sinks" for sediment brought in by rivers and these will accumulate as distinct but very thin (<1 cm) layers in the deeper parts of the lakes. All rivers entering Georgian Bay are largely devoid of any sediment which explains the remarkable clarity of the waters; the single exception being the Nottawasaga River at Wasaga in southern Georgian Bay which carries lots of sand and silt. Most lake-floor sediments are said to be "varved" in that they record the changing seasons with slightly coarser-textured layers deposited in summer and thinner, darker mud layers in winter when the lake is covered by ice. Pollen and dead micro-organisms fall to the lake floor and provide much information as to the climate and water conditions at the time. These can be cored from a boat or from ice and their contents studied in a laboratory. Successive layers can be radiocarbon dated. Geologists will also use geophysical methods to "see" the layers and map them over larger areas. One commonly used technique is seismic profiling which employs high frequency sound waves (like a fish finder) that are carefully tuned to penetrate the lake floor and reflect back from different sediment layers below thereby providing a "sub-bottom reflection profile." Work on many lakes in Ontario reveals a consistent pattern of layers below their floors recording a common history over the last 12,000 years. Rising lake levels are recorded by thick packages of sediment (identified on the accompanying diagram) whereas falling lake levels are recorded by erosion surfaces that cut across underlying sediments. The ups and downs of lake levels recorded on the profiles as seen here from Parry Sound record broad swings in climate. It has also been appreciated that such layers record past earthquakes in the form of layers that have been disturbed by sudden shaking. Such layers also contain micro fossils that key information on past climates and ecosystems in Georgian Bay.*

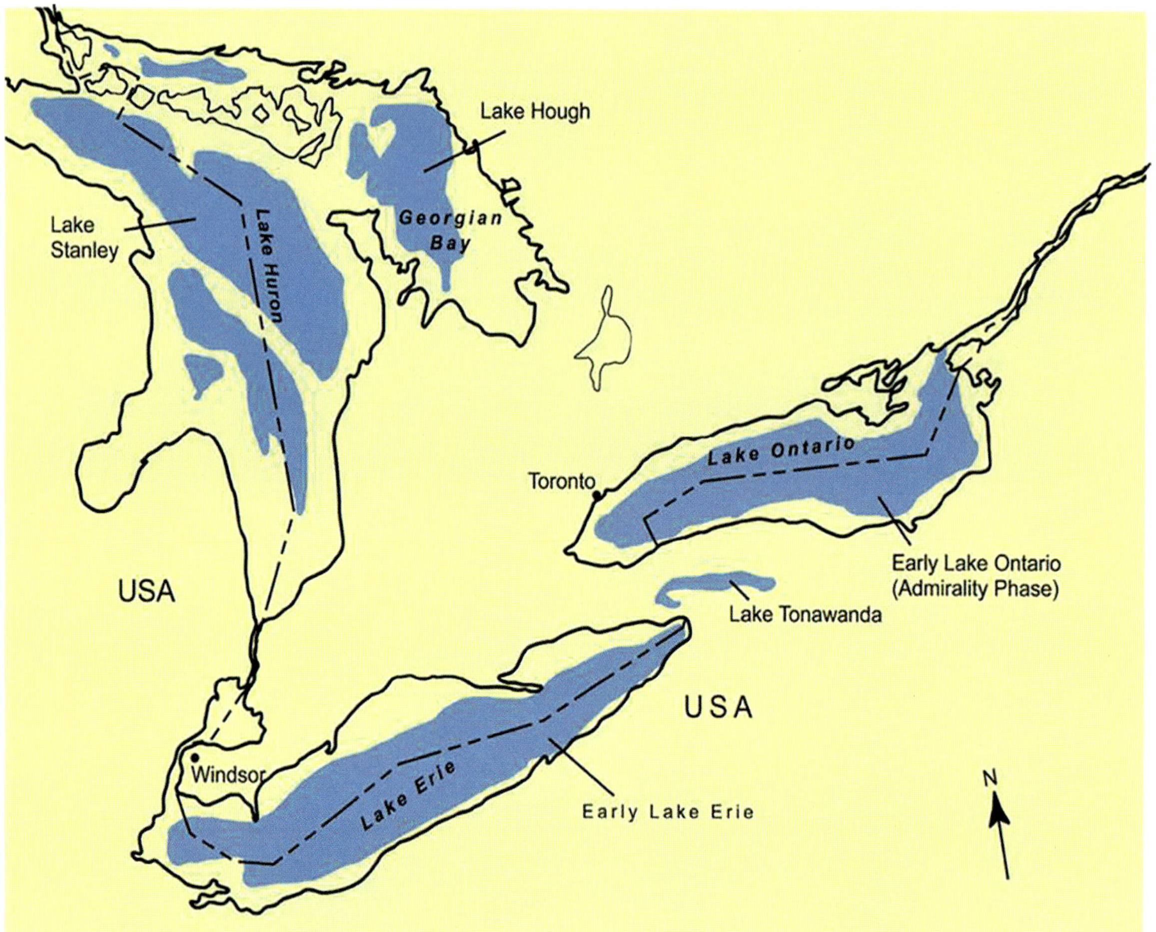

When Georgian Bay was dry

The Lesser Lakes of the early postglacial (c. 9000 to 5000 years): Lake Hough

With continued ice retreat, Lake Algonquin waters were able to escape eastward along the ice sheet's margin through what is now North Bay and thence through Mattawa, down the Ottawa River into the St. Lawrence Valley, where it was then flooded by the Champlain Sea. Other waters from glacial Lake Barlow also flowed down through the Ottawa. But this was temporary. Remember that the easternmost part of Ontario had been pushed down by glacio-isostatic depression under the great weight of the ice sheet much more than areas to the west so the land surface, upon being freed of ice, was tilted eastward. This allowed glacial lakes Algonquin and Iroquois to drain completely through the St. Lawrence River Valley when the ice finally retreated out of the region some 9,000 years ago. There was quite literally, nothing to hold back the waters of the lakes. This remarkable event is recorded by a long episode when much of what is now the floors of Lake Huron and Georgian Bay were exposed; these early postglacial "lesser lakes" are given the names Lake Hough in Georgian Bay and Lake Stanley in the Huron Basin. The work on the microfossils in Georgian Bay bottom sediments by Dr. Francine McCarthy at Brock University and her colleagues, has demonstrated that this interval was also associated with one of the driest phases of the last 10,000 years. This period is broadly called the *Hypsithermal* episode. There is good evidence that each of the Great Lake basins was closed with none or little outflow from one to the other. This was the time when Niagara Falls almost went dry because there was little outflow from Lake Erie. It has been estimated that snowfall was reduced to a third of modern values and the climate was much more continental with drier winters and hotter summers, something akin to modern-day Minnesota. Waters of the lesser lakes were also slightly brackish confirming aboriginal stories that the lakes were filled with "tears." This may be a predictor of the future of the Great Lakes in a warming world.

The last five thousand years: the crust rebounds and the Great Lakes recover

Wetter cooler conditions that begin after 5,000 years ago were part of a global climate trend that is referred to broadly as the *Neoglacial* episode. This was a time when glaciers once more began to flourish in the Canadian Rockies and elsewhere. Broadly speaking, the *Neoglacial* was wetter and cooler. Eastern Ontario was also rebounding fast bringing the land surface back to its pre-glacial level. As a result, water that had formerly leaked out of Georgian Bay through the French River and North Bay east along the Ottawa River was once again diverted back into the Huron Basin. This change essentially established the modern Great Lakes that now drain through the Niagara River. Trees that had grown across the now dry lake floors of the Hough and Stanley lake basins were drowned as lake levels rose. Accordingly, many Paleo-Indian archeological sites await discovery below the cool clear waters of Georgian Bay. Altogether, the modern Great Lakes as we know them today were essentially established 4,000 years ago.

How deep is Georgian Bay?

The surface of Georgian Bay lies at a mean elevation of 176 metres above sea level, only 3 metres above that of Lake Erie, but 100 metres above Lake Ontario. Whereas the bathymetry (depth) of Georgian Bay is well known, its submerged geology is not. The most striking feature of its floor is the presence of numerous deep channels, notably the Collingwood Channel, Thornbury Basin, and Owen Channel, and those along the eastern side of the Bruce Peninsula. The eastern channels, such as Colpoy's Bay and Owen Sound, terminate inland as narrow V-shaped "valleys" cut deep into the Niagara Escarpment. Indeed, the deepest water anywhere in Georgian Bay (150 m) occurs immediately offshore from Tobermory in two deep channels of the Bruce Basin that end abruptly at their western ends. Most of the channels on the Bay floor originate to the north just offshore the Thirty Thousand Islands where they are cut into the northernmost edge of the Paleozoic limestones and other sedimentary rocks which cover the Shield. These are known to geologists as "through valleys"' and are a common feature of the Shield/Paleozoic boundary further east in central Ontario.

It has been suggested that the heads of the channels in the Bruce Basin near Tobermory are actually drowned waterfalls akin to modern-day Niagara Falls. Indeed, in Anishinaabe legend this area is known as the "Place of Thunder Water." As we know, at the end of each ice age, large ice-dammed lakes held up in the Superior and Huron basins drained eastward across the Niagara Escarpment through the French River area and modern-day Lake Nipissing eastward to the Ottawa Valley. Certainly, the distinctive straightened margins of the channels near Tobermory have a more than passing resemblance to Niagara Gorge which has been cut by the steady upstream retreat of Niagara Falls over the past 10,000 years. Viewed in their entirety, however, the channels on the floor of Georgian Bay form a confused pattern unlike any simple river-cut channel system and this is because they are in the main, a relict ice-age landscape formed when Georgian Bay lay under glacier ice several kilometres thick.

On the eastern side of the Bay, Parry Sound (the "Big Sound") is the deepest freshwater port in the world, a glacially-carved basin that in places reaches more than 100 metres in depth. This area was surveyed and named by Captain Henry Bayfield in honour of Arctic explorer Sir William Edward Parry. The other area of relatively deep

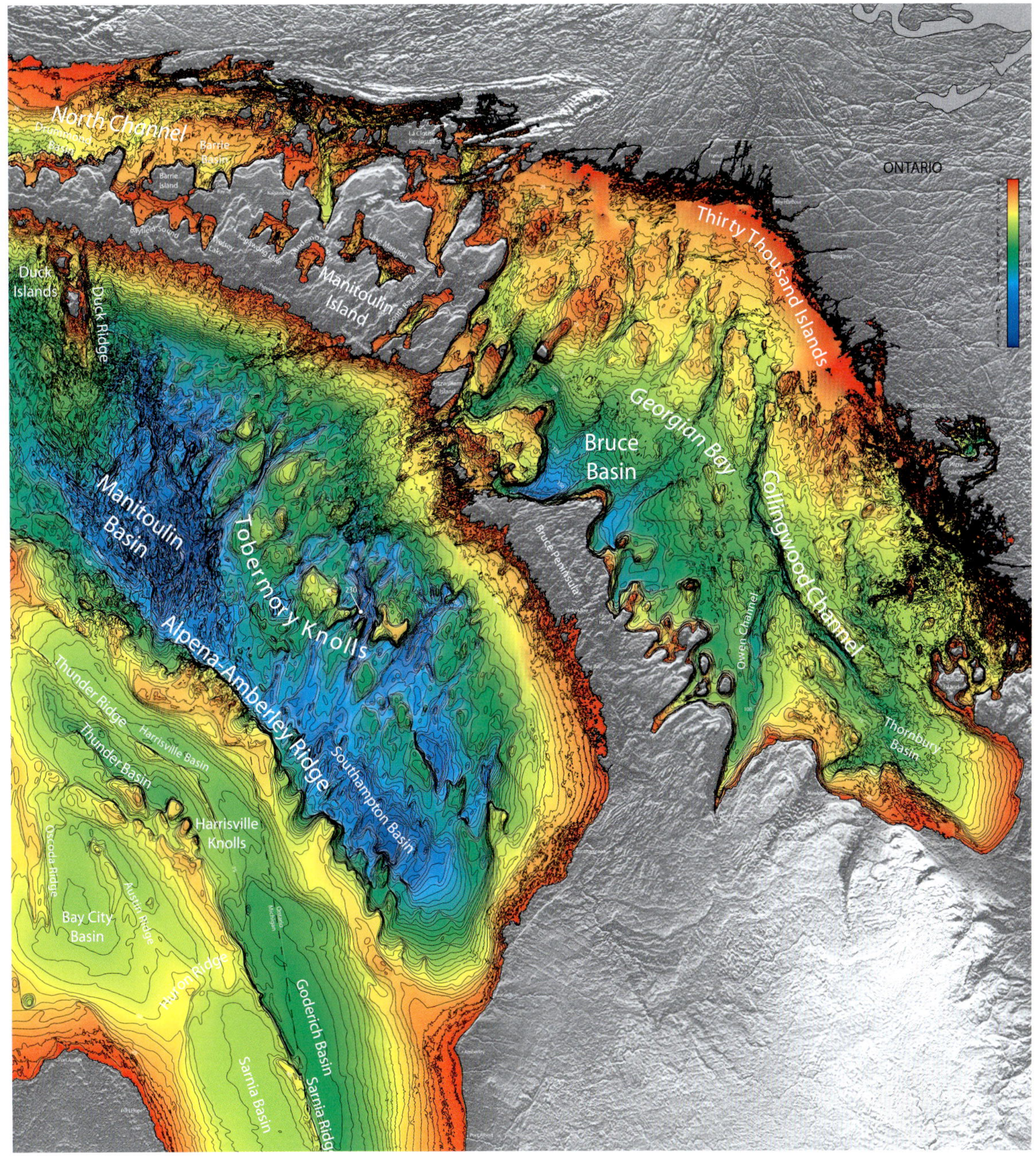

The floor of Georgian Bay in deep water is covered by thick mud and geologists and others have cored these sediments to extract pollen and learn of the changing climates and lake levels of the past 10,000 years. Incredibly, this work shows that the floor of Georgian Bay was exposed dry land in the recent past (between 9,000 and 5,000 years ago!) at a time of drier climate and much smaller "Great" Lakes whose surface elevations were some 30 metres less than is the case today. Much of the floor of Georgian Bay was also treed at that time only to be drowned as lake levels slowly recovered to the present day. Courtesy National Oceanic and Atmospheric Administrtion of the US.

Tobermory *lies at the northern tip of the Bruce Peninsula (representing the northern extension of the Niagara Peninsula) and was named in 1882 after its resemblance to the port on the Isle of Mull in Scotland. It has a remote setting, connected to Owen Sound in the south by a single road, Highway 6, and north by a seasonal passenger-car ferry, the 110 metre long MS Chi-Cheemaun (Ojibwe for "big canoe") which was built in Collingwood in 1974. This makes for a unique maritime setting matched only by Killarney on the eastern side of the Bay. Tobermory is widely known as the "freshwater scuba capital of the world" in reference to the clarity of the surrounding waters and the number of shipwrecks in the area. Water clarity is a reflection of the surrounding geology which is dominated by relatively hard Paleozoic dolostones and limestones; while these rocks dissolve they do not produce the muddy sediment caused by the weathering and breakdown of many other sedimentary rocks. Neither is there any large river to rework glacial sediments to the Bay.*

water occurs in Severn Sound in the far southeast corner of Georgian Bay. However, its enclosed waters have limited interchange with the Bay resulting in the need to protect lake water quality from nutrients carried by urban and agricultural runoff from surrounding communities such as Midland, Port McNicoll, and Penetanguishene. Phosphorus enrichment has in the past resulted in excessive algae growth and low oxygen conditions. As a result, the Sound was designated an Area of

Concern (AOC) in 1985 but with implementation of control measures, there has been much improvement and the area was delisted in 2003. It is here that the Trent-Severn Waterway, completed in 1920, enters Georgian Bay at Port Severn, 386 kilometres from its eastern end at Trenton, on Lake Ontario. A marine railway at Big Chute lifts boats some 18 metres up and over a large rock step in the Severn River. It is the only such feature in use today in North America.

Water circulation in the Bay

The surface area of Georgian Bay is 15,000 km^2 whereas that of its surrounding watershed is three times as large. Most of the water circulation in Georgian Bay is wind-driven with cool water moving into the Bay from Lake Superior via the North Channel at Little Current, and also from Lake Huron via the Main Channel between Manitoulin Island and the Bruce Peninsula. During summer (and less so in winter) prevailing westerly winds push surface waters from Lake Huron into the southern part of the Bay and set up an anti-clockwise rotation (forming what is called a "gyre") where water then flows north through the Thirty Thousand Islands and thence leaves the Bay for Lake Huron. Water remains in Georgian Bay on average for about eight years ("residence time") before being flushed out by return flows to Lake Huron through the Main Channel.

In summer there is a thin layer of warm water with a temperature of about 20°C that lies above the cold, deeper waters of Georgian Bay. This surface layer (which is called the "*epilimnion*") is usually only 15 metres thick and is lighter in density than the cold 4° C water that makes up most of the waters of Georgian Bay. Summer heating of the myriad islands and rocky shoals of the Thirty Thousand Islands plays a significant role in warming surrounding waters. It is not surprising that the warmest water temperatures tend to occur along the eastern side of the Bay. The coldest waters in summer generally occur along the opposite western shore of the Bay along the Bruce Peninsula in response to upwelling of cold water (< 4° C) that replaces warm surface waters of the epilimnion pushed eastward by winds. This results in rapid changes in water temperature enough to catch

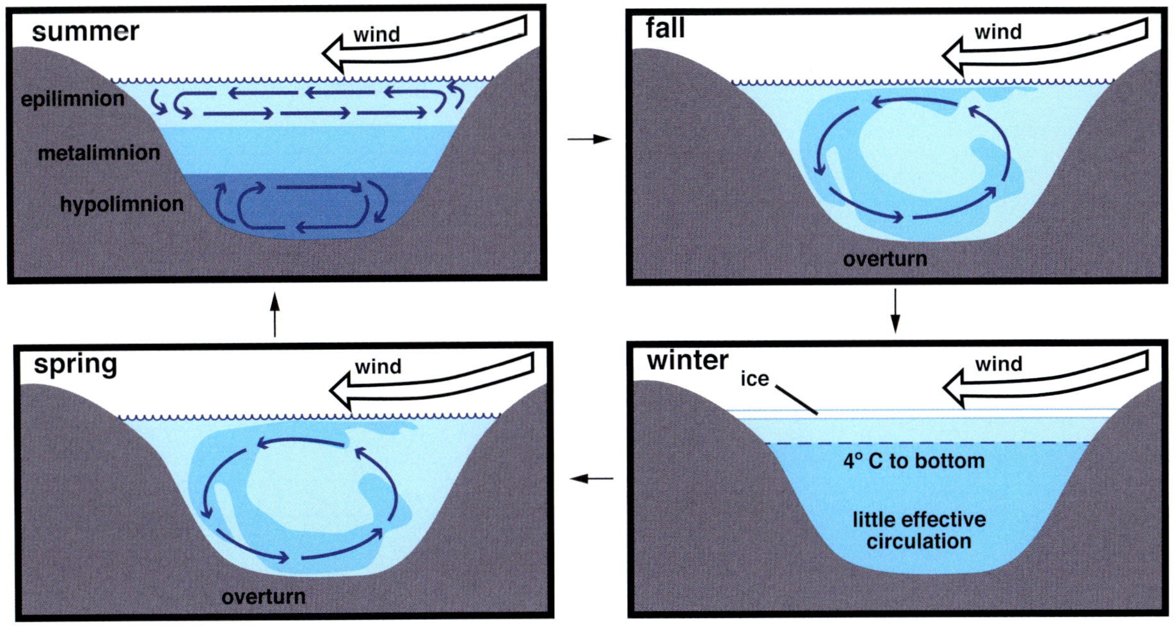

As surface waters cool in the fall, *they become dense and become completely mixed with underlying waters (the lake waters are said to "overturn") such that the lake is no longer stratified. Winter cooling of surface water re-establishes a lid of less dense water. This controls the seasonal movement of fish, especially lake trout who move into deeper cooler water in summer.*

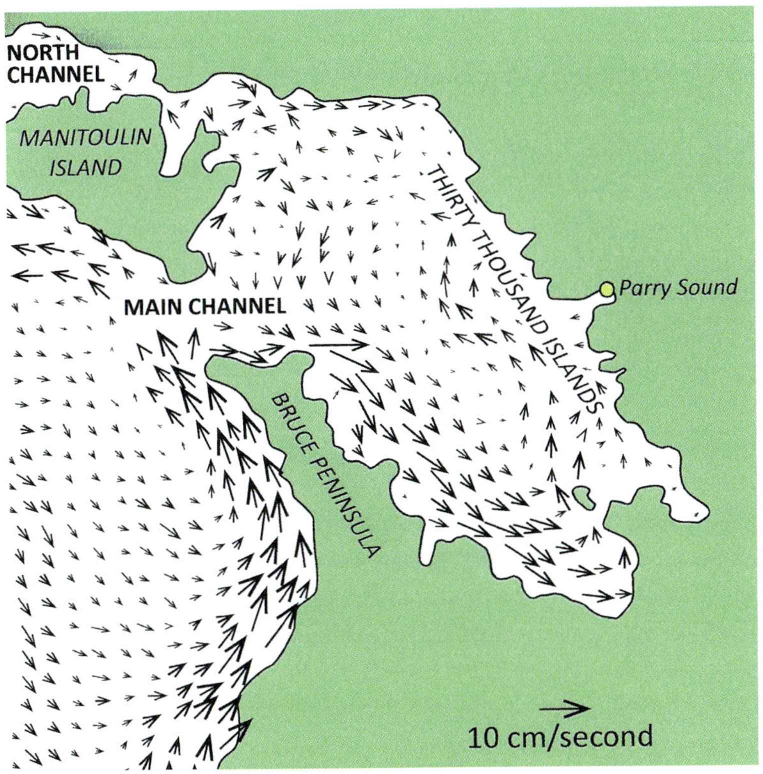

Wind driven water circulation *in summer in Georgian Bay. Note the presence of a large anticlockwise gyre.*

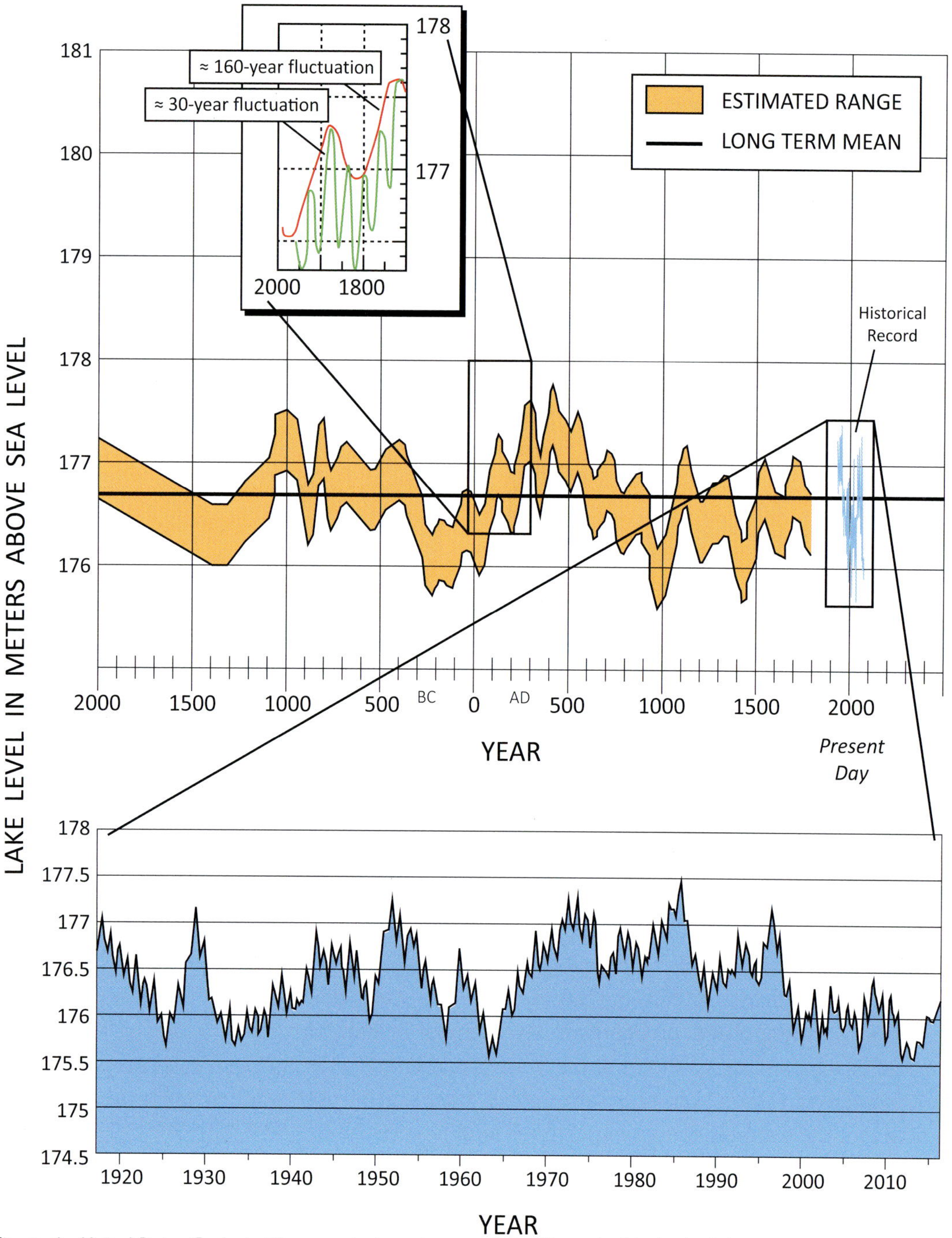

***According to the United States Geological Survey** and others, shorter periods of fluctuating lake levels of about 2 metres in magnitude, are superposed on longer term 30 and 100 year cycles giving rise to extremely low levels such as in 1926, the mid-1930s and mid-1960s, and very high levels such as in 1952, 1973, and 1985-86.*

swimmers unaware. The work of Dr. Mathew Wells at the University of Toronto has shown that the boundary between the epilimnion and underlying denser and much colder layers goes up and down by as much as 15 metres in the form of wave-like oscillations that propagate across the bay.

Winter freeze-up proceeds in the North Channel and along the shallow protected waters of the Thirty Thousand Islands in mid-December and proceeds lakeward to reach a maximum extent by mid-February; winters vary and so does maximum ice cover extent ranging from 25% to 95% of the Bay. Breakup starts in March and the Bay is normally free of ice by mid-April.

The waters of Lake Huron and Georgian Bay have warmed very slightly (0.048 °C/y) over the past 30 years largely in response to climate warming (since the cool wet 1970s) and the progressive shrinkage of ice covers from one winter to another. Another contributing factor is the earlier onset of summer stratification with depth (where relatively low density waters of the epilimnion, rests on deeper more dense waters of the hypolimnion) which in turn, creates a longer time frame for the warming of surface waters during any one summer.

Climatically driven cycles in lake level

And what of lake level changes in the past 4,000 years? Here we need to carefully differentiate between *absolute* changes in lake level and *relative* changes. Absolute changes in the level of Georgian Bay affect all parts of the lake's shoreline simultaneously and result from fluctuations in the amount of water stored within Lake Huron. These changes are driven by climate. Boaters and cottagers are very familiar with short-term seasonal changes in absolute level; the lake rises (to August) and falls thereafter to a low (in December) with a mean annual range of about 40 centimetres.

Over the last 4,000 years, longer term lake-level records stored in lake bottom sediments history reveals high stands associated with cooler climatic conditions and lows with warm climate periods. Likely causes are changes in the internal dynamics of the ocean-atmosphere system over the Pacific, externally driven changes to the ocean-atmosphere system created by variability in solar radiation, or volcanic activity and varying amounts of volcanic dust in the Earth's atmosphere. According to the United States Geological Survey, the water-level history of Lake Huron and Georgian Bay over the past

The Georgian Bay coastline is always evolving. In the long term, as the northern and eastern shores of Georgian Bay are being slowly uplifted following the last ice age, so its many shallow embayments and channels will be transformed into coastal marshes.

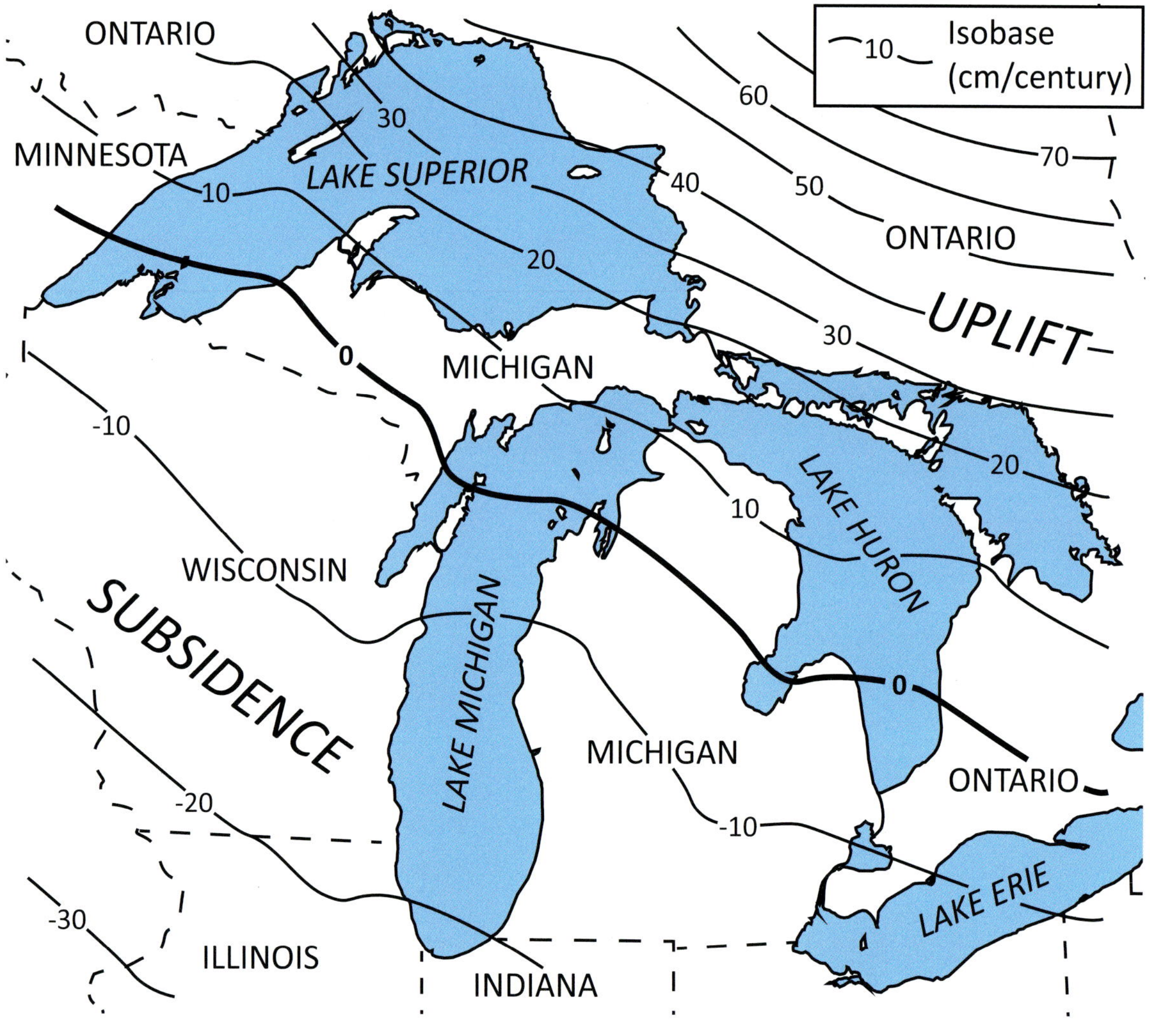

Forced down by the weight *of the last great ice sheet the crust of northern Georgian Bay is being uplifted between 20 and 30 centimetres a century; this is several times faster than that of locations to the south which results in slow tilting of the basin over time.*

4,000 years shows several roughly thousand-year-long phases of high water from 0 to 800, 1,100 to 2,000, and 2,300 to 3,300 years ago alternating with several low stands. These so-called "grand cycles" record a total fluctuation in water level of about 2 metres. There is also much discussion about a possible 160-year cycle and smaller cycles of about 30 years superposed on these grand cycles. Much more work is needed to more fully resolve the history of the lakes.

What is clear is that lake levels change very quickly just as we have seen from the low lows of 2013 (and interpreted by some as human-made and needing political interference; e.g., the "Stop the Drop" campaign) to the recent highs of 2014 to 2016 following

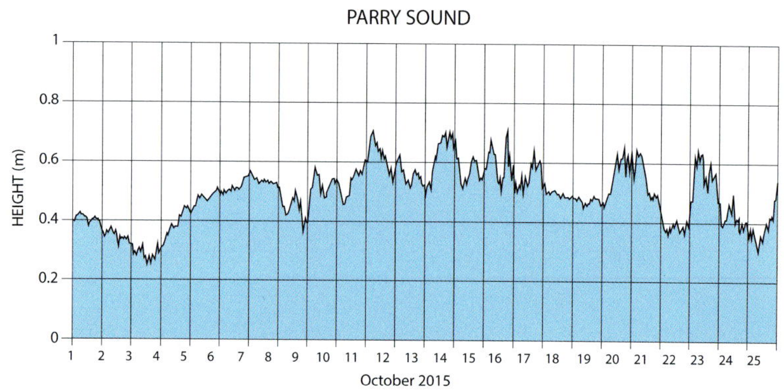

DAILY FLUCTUATIONS IN WATER LEVEL

relatively severe winters of greater than normal lake-ice cover and snow covers inland. Ice covers are key to stopping evaporation and water loss caused by the very cold, dry polar air masses that flow south across the lakes. The well-known *lake effect snowfalls* are the expression of huge volumes of water vapour being sucked off open water in winter and dumped downwind as snow. This is why lake levels are at their lowest at the end of winter before recovering when snow melts in spring.

Finally, there are small-scale irregular, daily and hourly changes in the level of Georgian Bay (and the other lakes) that are akin in many ways to tides on oceans and seas. No true lunar tides occur on the Bay but there are small-scale tide-like movements created by wind and changing ocean pressure as fronts move across the region. Fisheries and Oceans Canada keeps real-time records of these fluctuations which are of the order of about 25 centimetres. The record for October 2015 for Parry Sound is typical. These movements are called "seiches" (pronounced *say shhhh* and thought to originate from an old Swiss word meaning to "sway back and forth;" it is also referred to as "wind set-up") and was noted by the voyageurs because they created strong currents between islands and covered dangerous rocks. Early Native legends have it that these short-lived but very noticeable ups and downs of lake level are the "pulsing heart" of Georgian Bay.

The effects of postglacial rebound on lake levels; the legacy of the last ice age

Any discussion of lake level change also has to take into account *relative* lake level changes. This term refers to variations in water depth that are created not by climate but by vertical movements of the underlying basin of Georgian Bay. This is in response to rebound of the underlying Earth's crust following the last ice age. Imagine a bath tub that is slowly being tilted such that water level goes down on one side, and up on the other. The greatest rate of crustal uplift (some 21 cm a century) occurs along the northern shoreline of Georgian Bay near the outlet of the French River and along the North Channel near Manitoulin Island. This is simply because the last ice sheet was thicker in that area and so the amount of crustal subsidence under its enormous weight is greater. The coastline at Parry Sound is rising by about 21 centimetres every century; Tobermory by about 15 centimetres. In contrast, the southern shore of Georgian Bay at Collingwood is only rising at 10 centimetres a century and the outlet of Lake Huron, still further south at Sarnia, is actually *subsiding* because crustal rebound has ended in this area and the whole of the Huron Basin is being tilted down to the southwest. This means that over the last 50 years alone, most of Georgian Bay has risen at least 10 centimetres relative to the outlet of Lake Huron. As time goes by water depths near Parry Sound will continue to drop by 20 centimetres a century–independent of climatically-driven fluctuations in lake level–as crustal uplift continues. Hence the term *relative* lake-level change because the effects of glacio-isostatic rebound vary across Georgian Bay. Changes in relative lake level across Georgian Bay is clearly seen around the northern coast of Georgian Bay in the form of uplifted "high water marks" on cliffs which at first sight, suggest much higher lake levels in the past. But this is not so. The rocks on which the water marks were left have been uplifted further and further above the reach of modern high lake levels. In the long term, the entire Great Lakes basin will continue to be tilted down to the south as a consequence of glacio-isostatic rebound. This means that Georgian Bay will continue to rise quicker than those areas to the south for hundreds of years to come.

Human-made vs natural changes in lake level

The question always arises as to the significance of human activity on lake level changes in Georgian Bay. Modification of the St. Clair River outlet of Lake Huron as a result of dredging for gravel in the early 1900s and further deepening of navigation channels in the early 1960s are frequently cited as the main culprits behind recently lowered lake levels. This is expressed by widespread public concern aimed

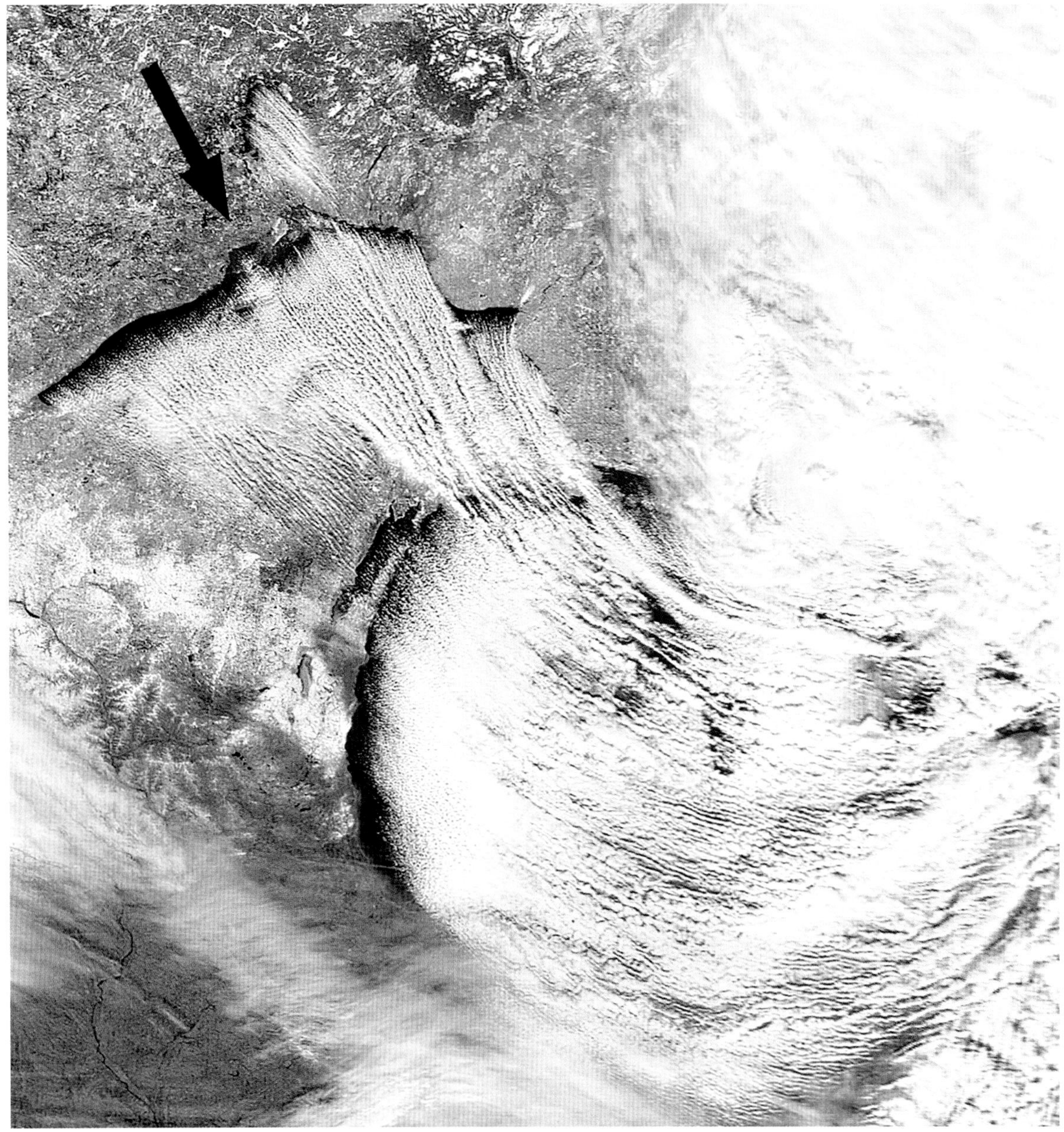

Cold arctic air flowing *across open lakes (arrowed) takes up enormous volumes of water vapour which is then dropped downwind as snow in the lee of the lakes. This gives rise to distinct "snowbelts" that occur on the east of Georgian Bay and at the east end of Lake Erie. Lake effect snowfall generates the all too familiar bands and streamers of snow that bedevil winter driving in the form of sudden whiteouts and very heavy snowfalls. The east and southeast side of Georgian Bay experiences frequent streamers as much as 50 kilometres wide and 150 kilometres in length which are capable of dumping as much as 25 centimetres of snow every hour downwind. Some studies indicate that lake effect snow controls the makeup of tree species in these areas and the winter movements of deer. Courtesy of National Oceanographic and Aeronautical Administration.*

Old water-level marks *(arrowed) near the mouth of Collins Inlet occur as much as 3 metres above the modern lake level. These have been uplifted as the crust of northern Georgian Bay slowly rebounds after the last ice age when ice as much as 2 kilometres thick covered the region. At first sight they seem to indicate higher water levels in the past but they provide evidence of glacio-isostatic rebound.*

at implementing remedial measures to somehow "Stop the Drop" in lake levels seen since 2000 (which are actually rapidly recovering as this chapter is being written). Dredging activities along the rivers are designed to accommodate the passage of larger ships and are estimated to have resulted in a total water level drop of about 40 centimetres over the last 50 years. This is about the same as naturally occurring seasonal fluctuations from summer to winter. While the lake level has declined it has not affected the natural seasonal and climate-driven rhythm in lake levels; the same 30-year periodicity is seen since 1960 and phases of *higher* water levels (such as the near record highs of the 1980s and the late 2010's) in Georgian Bay have also continued well after dredging along the St. Clair River. This underscores the overall importance of short-lived climatic factors in controlling lake levels, but the effects of dredging and continuing erosion along the St. Clair River are clearly of concern given that this allows more water to leave Lake Huron.

Conclusions

There are several conclusions that can be drawn regarding what is known of the processes that control lake levels in Georgian Bay and Lake Huron. Long-term geological studies tell us that natural variation in lake levels over the past 10,000 years, long before any human influence, were far greater than anything seen since 1860 when records began. Short term, small-scale climatically-driven variations are ongoing and their effects are superposed on the effects of regional postglacial rebound and the tilting of the Huron Basin, and of course a small decline in lake level as a consequence of dredging. The message here is that *change is constant.*

The other message which is especially relevant for conservation efforts around the Great Lakes and Georgian Bay, is that *healthy ecosystems depend on natural change.* Water levels fluctuate very rapidly in Georgian Bay on varying timescales; times of low level are quickly followed by high levels and vice versa; indeed, a wealth of studies reveal that rapid almost tide-like variation from season to season, day to day, and sometimes hour to hour, is a key to healthy lakeshore ecosystems. Great Lake coastal wetlands are diverse, highly dynamic and have evolved over thousands of years against a background of frequent changes in lake level. Ecosystem variability and resilience is known to be much reduced along the shorelines of those lakes (Lake Ontario) where water levels are regulated.

By demanding regulation of lake levels and the suppression of longer term changes, we place shoreline ecosystems at risk. As users and protectors of Georgian Bay we must adapt to such changes, not seek to control them at greater cost to the wider environment and its organisms.

What does the future have in store?

And what of future human-made climate changes and their effects on lake level in Georgian Bay? The jury is still out as to whether future changes will result in higher or lower lake levels because no one knows if future climate will be wetter or drier. Future model predictions are of limited practical use because they are speculative and entirely dependent on the assumptions made in regard to future climates (warmer and wetter or warmer and drier? Will there be more snow or less?). Some have argued that the low lake levels about 7,000 years ago, at the time of Lake Hough in Georgian Bay, might be a harbinger of things to come.

That said, what we do know is that lake levels will continue to fluctuate. Much of the marina and shoreline infrastructure around Georgian Bay was constructed at times of relative high lake levels and so is unprepared for episodic lower levels. The lesson here is that we need to develop shorelines in ways that respond to the ups and downs of lake levels using the last 100 years of water level history as a predictive guide. Fundamentally, these changes reflect fluctuations in the volumes of freshwater in the Great Lakes, a major natural resource shared by two countries. These lakes contain 20% of the world's surface freshwater. It is shocking how little we yet know of their history and their future. Going forward, conservation efforts must address both the *waters* and *land* of Georgian Bay.

(Courtesy Adobe Stock)

CHAPTER

5

LARGE MAMMALS OF GEORGIAN BAY

SUSAN HANNON

The Human Imprint

The history of large mammals in Georgian Bay is inextricably tied to that of the forest industry and logging. One hundred and fifty years ago, land surveyors were employed by the government to survey townships and other legal boundaries prior to settlement. They recorded details of forest composition along survey lines. These records have been compared to current forest inventory to see how the forest has changed. In the Great Lakes-St Lawrence forest region, deciduous trees have increased, particularly maple, poplar, and oak, while conifers (hemlock, red and white pine, spruce, larch, and balsam fir) have declined.

The lands to the east of Georgian Bay are part of the Great Lakes-St. Lawrence forest region, a transitional ecosystem that contains elements from the southern deciduous forests and the conifer-dominated boreal forest to the north. The deciduous species create our glorious fall colours: sugar maple (*Acer saccharum*), red maple (*A. rubrum*), northern red oak (*Quercus rubra*), beech (*Fagus grandifolia*), trembling aspen (*Populus tremuloides*), balsam poplar (*P. balsamifera*), and yellow (*Betula alleghaniensis*) and white birch (*B. papyrifera*). Mixed in with these are the conifers: white (*Pinus strobus*) and red pine (*Pinus resinosa*), eastern hemlock (*Tsuga canadensis*), white (*Picea glauca*) and black spruce (*P. Mariana*), balsam fir (*Abies balsamea*), jack pine (*P. banksiana*), and in the lowlands, black spruce, white cedar (*Thuja occidentalis*) and tamarack (*Larix laricina*).

The entry of European settlers and loggers into the Georgian Bay region opened up the forest, favouring animal species that fed on shrubs in openings and at forest edges. They also killed predators on their livestock and species that competed for food with domestic stock, or those who foraged on the crops eked out of the unforgiving landscape. Eastern elk and cougar disappeared entirely and woodland caribou, gray wolf, and moose were pushed farther north into the boreal forest. White-tailed deer, eastern wolves, coyotes, and bobcats moved up from the south. Bears and moose are more abundant than they once were, while lynx numbers are probably lower. While our views of predators and competitors are more enlightened today, we continue to shape changes in wildlife communities as we favour species that we can hunt and trap while we restore others, and alter habitats by suppressing wild fires, logging, building more roads and cottages and expanding urban areas.

The Moose *(Alces alces)*

Few species are so emblematic of the northern forests surrounding Georgian Bay as the moose, the largest member of the deer family (*Cervidae*) that includes elk (wapiti), white-tailed deer, and caribou. Standing up to 2 metres high at the shoulder, and sporting palmate antlers up to 180 centimetres across and a shaggy bell on its throat, a bull moose looks like some prehistoric holdover from the ice age. Moose bulls weigh up to 550 kilograms, cows up to 535 kilograms. While formidable looking, moose are nonetheless elusive and not often aggressive, except during the rut or when protecting a calf.

With their long legs and splayed feet, moose are well adapted to the Canadian Shield surrounding Georgian Bay and its many lakes and bogs. They feed with alacrity on land and in water, and are able to dive up to 5 metres below the surface to rip up aquatic plants. Fatty pads and specialized muscles in the nose enable moose to close their nostrils underwater.

After the long winter, moose have depleted their fat stores and are in poor condition. In spring they feed on leaves and shoots of shrubs until aquatic plants are available. Then they may move up to 30 kilometres to the best aquatic feeding areas in beaver ponds, wetlands, creeks, and shallow weedy bays. Aquatic plants provide important nutrients (especially sodium) to the diet. At this time, moose may also frequent mineral licks, areas where salt and minerals seep out of the ground. But they are also attracted to salty pools formed from runoff from roads where they can be hit by vehicles.

Moose are large animals with dark coats and can overheat in summer. To avoid heat stress they bed in cool, shady areas close to their summer feeding areas. These bedding areas are found in mature hardwood or conifer stands with high canopy cover in lowland areas. In late summer, fall, and early winter moose move to shrubby areas of the upland forests where they feed on leaves and twigs of shrubs such as willow, aspen, red-osier dogwood, maple, birch, cherry, beaked hazelnut, and balsam fir. Spring and summer feeding sites are very important for moose so they can build up their fat supplies for the rut and for the long winter ahead.

In winter, life is harder as snow depths increase and moose must browse on the twigs of shrubs, harder to digest and lower in protein than leaves. They must also restrict their movements to save energy. If food becomes scarce they may resort to stripping the bark from aspen trees. A moose can store up to 45 kilograms of food in its stomach, which is later digested when the animal is resting: regurgitating and rechewing the fermented food.

Young moose calves are very vulnerable to predation by wolves and bear. Their main protection is not camouflage, but their aggressive mothers. Wolves hunt in packs and "test" moose to see if they are healthy: an adult moose in good condition can wound or kill a wolf. Only about 1 in 12 wolf-moose encounters end with the wolves successfully killing a moose. Wolves have the advantage in deep snow with a crust on it or on ice as they can move quickly to overtake a moose that may be bogged down.

Moose are also susceptible to ticks and may lose a lot of blood, or because of the associated itch will rub off their brown outer hair; these "ghost moose" rarely survive the winter. Where moose and deer ranges overlap, moose can become infected with a meningeal worm *Parelaphostrongylus tenuis* that destroys brain tissues and causes a condition called "circling disease" where the animal stumbles around aimlessly. The parasite is picked up by the moose during feeding: it infects snails or slugs that have been in contact with infected deer feces. White-tailed deer are adapted to the parasite, but in moose it is often fatal.

Moose thrive in upland areas, where young deciduous stands with an abundance of shrubs and young trees for food occur next to older forest with dense conifer cover for protection from the wind and deep snow. Protecting aquatic feeding areas, calving sites, and summer thermal cover is also critical to maintaining good moose populations. Forest management plans often delineate Moose Emphasis Areas where timber harvesting produces irregularly shaped cuts with scattered patches of dense mature conifer for winter shelter, and mature forest cover for summer thermal cover near foraging areas.

White-tailed deer (*Odocoileus virginianus*)

White-tailed deer are the most numerous of the large mammals in Georgian Bay country; if the moose is the linebacker of the cervid world, then the white-tailed deer is the ballerina. Their slender legs and dainty hooves belie an amazing athleticism. Deer can leap fences over two times their height, jump horizontally 9 metres, and can reach speeds of 75 km/hr with their tails waving like white flags signalling to predators that they have been detected. They are a familiar sight to cottagers and townsfolk while tolerating humans in their quest for food, often coming into gardens to feed. However, when the first Europeans arrived, white-tailed deer were only found in the most southern parts of Canada. The deer moved north with the settlers as the forests were cut and burned and crops planted, transforming the landscape into a patchwork of forest, shrubby areas, and open fields, a deer paradise.

Bucks and does normally spend the spring and summer apart, foraging to regain body condition and build-up fat for the long winter ahead. Males begin to regrow antlers in spring. These have an internal and external (the velvet) blood supply. This velvet begins to peel off in late summer when the antlers have finished growing. The rubbing of the antlers on shrubs and trees to remove velvet leaves tell-tale scrapes on the bark. Antlers are shed after the rut.

In spring and summer, deer choose green vegetation from a variety of woody plants, grasses, herbs, and forbs, and make sorties into agricultural fields for clover and corn. While primarily a vegetarian, the white-tailed deer has been seen eating eggs and nestlings out of low-lying songbird nests. In late summer and fall, acorns and beech nuts are a favourite food, energy rich and contribute to a build-up of fat for winter. These mast crops are an important resource for many wildlife species.

As fall turns to winter and snow exceeds 50 centimetres deep, deer migrate from summer ranges to traditional winter ranges, called "deer yards" with dense conifer (hemlock, cedar) cover where snow depths are lower and there is protection from the cold and wind. The conifer branches capture the snow, much of which sublimates before it falls to the ground. Here they browse on woody twigs and buds of shrubs such as hazel, dogwood, yellow birch, and maple and the foliage of cedar and hemlock. The yards cover only 10-15% of the summer range, so deer are crowded. The larger deer yard in Georgian Bay area occurs south of Lake Nipissing in the Loring-Restoule area where some 10,000 deer spend the winter months.

Long winters with extreme cold and deep snow result in massive die-offs of deer. Fawns are particularly susceptible as they don't have the energy stores of adult deer and have difficulty eluding wolves in deep snow. Bucks coming out of the rut in poor condition may also be susceptible to malnutrition and pregnant does lose their fawns.

White-tailed deer

(Courtesy Adobe Stock)

Deer have a number of adaptations to conserve energy in winter. They move less and decrease their body temperature and metabolism. Their winter coats are 10 times thicker than the summer coat and are highly insulating. In long, severe winters, deer begin to break down their muscle tissues for energy and become gaunt and weak. Deer can lose up to 25% of their body weight without affecting survival, but at the end of long winters many have nothing left and mortality can be very high.

Feeding starving deer in winter is not recommended as it artificially inflates survival and increases deer numbers causing over browsing of the natural food supply. Feeding is a last resort after a long cold winter when deer are unlikely to survive. Deer cannot digest alfalfa, corn or hay which are too high in fibre and starch; high-protein deer pellets are favoured though the best way to help starving deer is to break trails through the snow from the yards to areas with good browsing using snowmobiles.

Fawns are the most susceptible segment of the large mammal population to predation and can be captured by wolves, coyotes, black bears, bobcats, and free-roaming dogs. Deer of all ages are killed by wolves, especially in late winter in the yards when they are weak and the snow crust allows wolves to move quickly, while the deer sink. In summer, many deer are killed on the roads or by predators and in the fall hunting mortality occurs. Deer move about at dawn and dusk and at night, which makes them susceptible to collisions with vehicles.

Landowners can also improve deer range on their properties by planting or protecting oaks and apple trees and seeding grass and clover into openings, such as old fields or logging roads. Cutting pole sized trees such as maple and poplar will encourage suckering to produce deer food and shrubs such as dogwood and viburnum can be planted.

Eastern Elk (*Cervus elaphus canadensis*)

The eastern elk, a subspecies of the American elk or wapiti, once thrived in Ontario, Quebec, and the eastern United States but became extinct in the late 1800s because of overhunting and a loss of their habitat to settlement, logging, and agriculture. The last known eastern elk in Ontario was shot in 1893 outside of North Bay. In the 1930s some western elk were introduced into Ontario but most were later destroyed because of concerns about parasites passing to domestic

Elk (Courtesy Adobe Stock)

stock. Between 1998 and 2001 over 400 elk from Elk Island National Park in Alberta were released at two sites in Georgian Bay country at the French River (near Burwash) and along the north shore of Lake Huron. They were equipped with radio collars, allowing biologists to monitor their fates. The primary causes of elk mortality in northern Ontario included wolf predation, illegal shooting (mostly hunters misidentifying elk), and collisions with trains.

Black bears (*Ursus americanus americanus*)

Black bears are a common denizen of the forests, meadows, and roadsides of Georgian Bay and unwelcome visitors to towns, villages, and cottages. While usually black, their fur can range from brown to cinnamon to blonde. Black bears are generalists living like deer in a wide range of habitats. They reach highest abundance in habitats that combine early successional forest, open areas, edge, and some mature forest. They are omnivorous and have adapted well to human changes on the landscape. Their numbers are increasing over much of Ontario.

In fact, about 80% of their diet is plants.

Bears emerging from their dens in spring face the challenge of revving up their metabolism from winter hibernation levels to spring and summer levels. During this period of two to three weeks, a phase referred to as "walking hibernation,"' bears eat and drink very little. As plants in meadows and road edges begin to green up, bears ramp up their feeding. The tender shoots and flowers on new growth are relatively easy to digest since the cellulose hasn't hardened yet. Bears also feed on the new leaves of trees, especially aspen: either bending down saplings to reach them or climbing the trees.

As the season progresses and plant cellulose hardens and becomes more indigestible, bears turn to insects, overturning decaying logs and rocks to find ants and raiding bee nests. They even feed on tent caterpillars when they are common.

As berries ripen and become rich in energy, they become an important food source, but any food encountered is fair game, including bird eggs. Bears are known to travel up to 100 kilometres from their home range to forage on blueberry patches that they foraged on as cubs. Other important berries include cherries, mountain ash, Saskatoon, cranberry, and gooseberry. Berry crops can be highly variable over years with fruit production reduced by late spring frosts and dry or cool summer weather.

As spring turns to summer, bears begin to search for mates. Black bears have fairly low reproductive rates for a large mammal, the lowest in North America with the exception of the musk ox. The home ranges of males are up to 10 times the area of female home ranges (20-25 km^2) and several females are encompassed in his area. A single female is in estrus for only about five days and hormones in her urine signal her reproductive state to the male. Bears have a fairly unique form of birth control, also found in weasels, seals, otters, bats, armadillos, kangaroos, nutrias, and red pandas and is called delayed implantation. The fertilized egg does not implant into the uterus immediately, but floats free for about four months until the female enters her den in early to mid-October. If food supply during the summer and autumn has been poor and the female has not attained a weight of at least 70 kilograms, then the fertilized egg is absorbed. This is an adaptation to ensure that cubs are not born when the female is unable to produce milk and also avoids wasted reproduction.

Cubs are born in the den during hibernation, a very unusual situation for mammals as most mammals give birth in spring when food is more abundant. But bears hibernate through winter and must accumulate enough fat stores through the rest of the year to trigger hibernation. Mating in the fall is too energetically costly as bears are feeding frantically to build up fat stores for hibernation and pregnancy during this time would put stress on the female. Thus implantation is delayed and cubs are born after only two months of gestation, weighing between 200-330 grams in early January, blind and toothless but development is rapid as they feed on their mother's fat and protein-rich milk. By the time they emerge in April, they have gained 20 times that much and are ready to face the rigours of life outside the den. Cubs stay with their mothers through their first year, denning with her in the fall and then leaving in May or June when they are 17 to 18 months old. Females tend to settle close to their mother's home range while males wander further afield.

In late summer or early fall, bears enter a period of *hyperphagia*, excessive eating and drinking, to fatten up for hibernation. Caloric intake can triple during this period. Aside from berries, the other important food source is the mast (nuts) from deciduous trees and shrubs: oaks, beech, and beaked hazelnut. Mast crops are notoriously variable from year to year, partly a result of weather conditions. Cold spring temperatures can kill flowers, reducing seed set and summer drought can hinder development of the nuts.

The seed set of berries and mast-producing deciduous trees such as oak have an important effect on bear populations. In years of poor crops, bears must stay out longer to feed in order to achieve their hibernation weight. They also move over larger areas in search of food, bringing them into contact with humans and their food sources and making them more susceptible to hunting and other human caused mortality.

Bears begin to search for dens in September and October. They can dig a burrow in the earth, or use hollow trees, rock crevices or sometimes crawl under brush piles or the roots of windblown trees, using a different den each year. Prior to entering dens, bears go through a period of reducing their metabolism in preparation for hibernation. They eat less, but continue drinking, heart rates fall and bears become lethargic. Dens are not well insulated and bears depend on their fat and dense body hair to keep them warm over the winter.

Currently, the density of black bears in Georgian Bay country

Black bear

(Courtesy Adobe Stock)

Lynx

(Courtesy Adobe Stock)

is about 48 bears per 100 km^2 and the population is regarded as healthy. Ontario boasts the second largest population of black bears in Canada, some 75,000-105,000 animals, second only to British Columbia. The current rate of bears harvested by hunters and trappers is about 6-8% of the population. In the years of European settlement, bears were perceived as threats to life and livelihood, and they were killed when encountered. A bounty was in place from 1793-1960. As the behaviour and ecological role of bears became better known and the economic benefits of a healthy bear population realized, they were classified as a big game species in Ontario and regulations were enacted. Compensation programs were brought in to reimburse farmers for losses of livestock and beehives. The spring and fall bear hunts became popular, particularly among non-resident hunters, and became an important economic driver in central and northern Ontario. In 1999, the MNRF cancelled the spring bear hunt to allay fears raised by some animal rights and environmental organizations that females were being shot and cubs orphaned. Today, the concern is that bear numbers are at unacceptable levels and kill rates on moose calves are too high, more livestock and crops have been depredated and public safety is compromised as more bears are coming into towns searching for food.

Cottagers and townsfolk have important roles to play by making their properties unappealing to bears by keeping properties clean. Reductions in human-bear conflicts can also be achieved through enhancing natural food supplies and reducing human food supplies that are accessible to bears. Logging opens up the understory of the forest allowing growth of the berry and nut producing shrubs that bears love. Important mast trees, such as red oak, need to be protected, and den sites avoided during hibernation and protected using no-cut buffers around them.

The Wild Cats of Georgian Bay

These are the most elusive mammals of Georgian Bay. If you are extremely lucky, you might spot a Canada lynx (*Lynx canadensis*), bobcat (*Lynx rufus*), or cougar (*Puma concolor couguar*). The southern Canadian Shield country marks the southern boundary range of the Canada lynx and the northern boundary of the bobcat. Often confused, the two lynx species have many similarities in their biology, can be competitors for food and territory, and occasionally hybridize. The cougar once roamed over most of North America but was exterminated from the east by the late 1880s, another casualty of the spread of European settlement. It is now making a tenuous comeback.

Telling the difference between a lynx and bobcat is not easy. The lynx is larger, has bigger furrier paws, a longer back foot, longer ear tufts, and shorter tail. Its back legs are quite a lot longer than the front giving it a sloped appearance which is not so noticeable in a bobcat. The bobcat has distinct spots and a reddish cast to its fur, where the lynx's coat is mostly gray with indistinct spots.

The Canada lynx experiences dramatic fluctuations in its population size and over a 10-year cycle, their abundance fluctuates 3 to 17 times depending on the abundance of its main prey, the snowshoe hare (*Lepus americanus*). These cycles were first noted by Europeans (and undoubtedly well known by Indigenous people) from trapping records maintained by the Hudson's Bay Company. These showed a striking cycle of ups and downs in snowshoe hares followed with a one to two year lag by lynx. When hare numbers crash, lynx mortality increases, reproductive rates decline, and dispersal increases.

Georgian Bay country is not prime habitat for lynx because of habitat loss and fragmentation. They are much more common in the boreal forest to the north, where hares are more common. A single lynx was trapped in the Parry Sound district between 2004 and 2013 compared with 11 bobcats and 311 coyotes.

Estimating densities of lynx is extremely difficult, but in the southern boreal forest there may be only 2-3 lynx/100 km^2. In Ontario the lynx is classified as a furbearer and trapping is the main anthropogenic cause of mortality. Lynx are relatively easy to trap and when fur prices are high many lynx are taken. Lynx numbers declined significantly in Canada between 1920 to 1950 due to over trapping and habitat loss. Numbers rebounded when fur prices decreased. In 2000, Ontario reported that its lynx population was secure and that lynx were common.

Cougars in Georgian Bay?

The last documented wild cougar in Ontario was shot near Creemore southeast of Collingwood in 1894. Putative sightings of cougars have been increasing, many west of Thunder Bay but lately more and more in central and Southern Ontario, including near Parry Sound and Muskoka. In 1971 the cougar was listed as an endangered species;

Cougar

(Courtesy Adobe Stock)

many recent sightings turn out to be of other species such as lynx, bobcat, coyote, and fisher but there is plenty of cougar habitat in Ontario and a large prey base of white-tailed deer (estimated at 400,000 to 500,000 animals). Cougars are habitat generalists and have amazing dispersal abilities: some radio-tagged animals in the western United States were documented to move over 1,000 kilometres. A growing population of cougars will present new challenges to farmers and other landowners–it will be interesting to see if we can coexist.

The Eastern Wolf

It would be hard to think of an animal that has engendered such a range of emotions in humans as the wolf. Interest in wolves is high among the general public. Harnessing this interest economically could help to keep larger landscapes intact. Ontario is home to three members of the Canid family: eastern (*Canis lycaon*) and gray wolves (*Canis lupus*), and the coyote (*Canis latrans*). The eastern wolf is smaller than the gray wolf that lives farther north in the boreal forest and tundra lands north of Lake Superior. Gray wolves were taken during the fur trade and further persecuted as settlers moved north with their livestock. Deforestation and conversion of the land to agriculture further reduced their numbers. The gray wolf was extirpated from Georgian Bay between 1850 and 1900. Today, the eastern wolf occurs on the southern margins of the Canadian Shield around Georgian Bay north and west to Lake Nipigon.

Coyotes are a relatively new addition to the fauna of Ontario, historically restricted to the plains, deserts, and parklands of central North America. They are the smallest of the Canid family. They are also highly adaptable and can tolerate living close to humans in disturbed habitats and agricultural lands. They are often killed by wolves when the two species overlap. Hence, as wolf numbers declined coyote numbers went up and they moved into Southern Ontario in the early 1900s. In areas where Canid species overlap, there may be some hybridization, but normally wolves will kill coyotes that trespass on their territory.

Eastern wolves are hybridization intermediaries: they have mated with gray wolves at the northern edge of their range and with invading coyotes to the south. The latter has given rise to an intermediate form called the Tweed wolf, eastern coyote, or *coywolf*. In the area between Parry Sound and Huntsville, which has much more human activity than in Algonquin Park for instance, most of the wild canids are coyote/wolf hybrids or eastern wolf/gray wolf hybrids.

About 99% of the diet of eastern wolves is white-tailed deer, taken mainly in winter, supplemented with beaver in the snow-free months and moose. Eastern wolves are smaller than gray wolves so are less likely to take larger ungulates such as moose or elk. As a top predator they help to keep prey populations in check and cull out weak individuals, old, physically impaired, or diseased. The presence of a wolf pack helps to move the prey around the landscape avoiding concentrations in habitats such as riparian areas. Wolf kills provide food to the many scavengers in the forest such as crows, ravens, bears, and jays.

The lifespan of a wolf in nature averages five or six years when they live in packs, facilitating group pursuit of prey which allows larger animals to be taken down. In the well-studied Algonquin Park, packs of eastern wolves comprise four to five animals, usually a breeding pair and offspring of previous years and patrol territories from 150-500 km^2 that they loosely defend from other packs. Territories are marked by scent and advertised to other wolves by howling. Mating occurs in mid to late February between the dominant adult pair in the pack. A den is excavated into sandy well-drained soil and four to seven pups are born in mid to late April. They are nursed for the first six to eight weeks and subsequently they are moved to rendezvous sites where they wait for the adults to return with food.

At the rendezvous sites which can be in a bog or beaver meadow, pups are left to their own devices while they play and hone their hunting skills on small creatures in the area. Pups may be moved several times, until they begin hunting with adults when they are about four months of age. The simple rule is that the higher the deer population the higher the survival rate; the more roads the lower the survival.

Admired by Aboriginal people, hated by settlers and hunters, and enshrined by others as a symbol of our last wild lands, the fortunes of the wolf have vacillated over history. Thousands of pelts were taken during the fur trade but when settlers arrived in Southern Ontario, the war on the wolf took on new dimensions. They were killed with impunity to prevent depredations of livestock and in Ontario in 1793 a bounty was introduced. The bounty was repealed in 1972 and wolves were managed under the Game and Fish Act. Currently, because of

Eastern wolf

(Courtesy Adobe Stock)

difficulties in distinguishing between eastern wolves and coyotes, in the primary range of the eastern wolf, wolves and coyotes are managed as if they were the same population. Hunting and trapping are allowed from 15 September – 31 March, with a bag limit of two wolves or coyotes per year for hunters and no bag limit for trappers on registered traplines. While not currently threatened or endangered in Ontario, the eastern wolf has been designated as a species of "Special Concern" meaning that a wildlife species may become threatened or endangered because of a combination of biological characteristics and identified threats. Future conservation efforts for wolves will have to rely on better census methods to more accurately determine population size and better estimates of how many wolves are harvested by hunters and in defence of property. As the human population in the south slowly expands north into Georgian Bay, intact wolf habitat is being lost. Roads provide improved access to hunters and cause road kills, increasing wolf mortality.

Nearly 90% of the 14,000 wildlife/vehicle collisions reported in Ontario each year occurs on two-lane roads outside of urban areas in good weather. Most occur at dusk or dawn. Collisions peak in May and June and again from October to January.

In 2012 MOT built a combined overpass and underpass near Burwash as part of the twinning of highway 69 from Parry Sound to Sudbury. They also fenced 10 kilometres of the roadway and installed some Texas gates and a wildlife path under a bridge. The overpass is similar to those built in Banff National Park and is planted with trees, shrubs, rock piles and brush so it looks natural. Wildlife cameras were set up at the overpass and underpass to record what species of wildlife were using the crossings. So far, moose, deer, black bear, wolves, coyotes, bobcats, and red fox have used the structures. Large mammals, except for coyotes, were more likely to use the overpass than the underpass.

Common Merganser

Photo: Randy Holland

CHAPTER

6

GEORGIAN BAY BIRDS

BOB WHITTAM

Barred owl

Photo: Randy Holland

Georgian Bay has an excellent representation of Ontario's birds with about 300 species recorded here of which some 170 breed in the area. Waterfowl are an important bird family followed by the loons and grebes which are the most primitive types of birds. Then, following in evolutionary order, are wading birds (herons, egrets), waterfowl (ducks, geese and swans), raptors (eagles, falcons, hawks), upland game birds (ruffed grouse), shorebirds, gulls and terns, owls, woodpeckers, flycatchers, jays and crows, swallows, chickadees, wrens, wood warblers, blackbirds and sparrows. What follows is a brief summation of the principal species around Georgian Bay.

The common loon (*Gavia immer*) is the provincial bird of Ontario appearing on the back of the $1 coin, and is so named from its ungainly gate on land (derived from the Icelandic *'lúinn'*) as its legs occur to the rear of its body; ideal for diving but not walking. With its large webbed feet it is an expert diver and swimmer going as deep as 60 metres where it can remain submerged for up to 3 minutes feeding on various minnows. It is ideally suited to Georgian Bay and its myriad waterways and lakes where it is a seasonal visitor arriving as soon as the ice goes out. It winters along the east coast and in early April, can be seen flying fast (75 km/h) high overhead prospecting for lakes with open water, often returning southwards if it has arrived too early. It is revered for its tremulous calls, hoots and yodels which are emblematic of the Canadian Shield country surrounding Georgian Bay.

Many bird species are at their northern limit in Georgian Bay such as cuckoos, wrens, warblers, red headed and red bellied woodpeckers. Those at their southern limit are northern birds such as Spruce Grouse and Gray jays. The Georgian Bay shoreline provides feeding and staging areas for migrating birds and some shoreline features are important for nesting. The sand flats and dunes at Wasaga Beach have attracted the endangered Piping Plover. Coastal marshes support waterfowl and rails, and the islands of Georgian Bay are important for nesting waterbirds such as Common Loon, Osprey, Common Merganser, Spotted Sandpiper, Double-crested Cormorant, Ring-billed Gull, Herring Gull, Common Tern, Caspian Tern and American Pipit. Aerial foragers such as the Purple Martin and the Bank Swallow find suitable habitat along the Severn Sound shoreline. Fire has greatly influenced bird populations by providing early successional habitat for Chestnut-sided Warbler, American

Kestrels and Sharp-tailed Grouse. Forest Tent Caterpillars and other defoliators provide an immediate and beneficial food source for warblers in northern parts of Georgian Bay while Tennessee Warblers and Bay-breasted Warblers eat Spruce Budworm larvae. The Black-billed Cuckoo eats tent caterpillars and is most abundant during outbreak years. Beaver ponds provide nesting habitat for Hooded Mergansers, beaver meadows for Sedge Wrens and shrubs that grow on the edge of beaver ponds attract Golden-winged Warbler and Alder Flycatcher. While there has been a recent decline in grassland birds and aerial foragers, there are overall increases in some forest birds, wetland birds and raptors. Atlas research found an increase in big birds in the province. Eleven out of twelve species of large birds including the Trumpeter Swan, Wild Turkey, Bald Eagle, Sandhill Crane, Canada Goose, Great Blue Heron, and Turkey Vulture all showed increases in the last 20 years. This is due in part to the decline in the use of DDT, and a reduction in the wanton shooting of these species.

Barred Owl

In the forests surrounding Georgian Bay, the most common owls are the Great Gray Owl (in northern sectors only), the Barred Owl, the Boreal Owl and the Northern Saw-whet Owl. The most abundant owl is the Barred Owl and its call is heard quite frequently in the late winter and early spring. By early April, it is in full chorus with its familiar 'who hoots for you, who hoots for you-all?' They live in and around Georgian Bay year round.

The demise of the Passenger Pigeon

The early Georgian Bay explorers would have seen an amazing amount of wildlife but one of the most impressive spectacles will never be witnessed again. On a sunny spring morning in 1876, a giant cloud of Passenger Pigeons obscured the sky over the Bay and still remained as the evening chorus of songbirds began. The cloud was a formation of millions of pigeons passing by and stopping on their migratory journey. In a relatively short time the population of Passenger Pigeons vanished. In 1605, Champlain recorded 'countless numbers.' The Jesuit Gabriel Sagard who lived with the Hurons from 1623-1624 made the earliest reference to the Passenger Pigeon in the Georgian Bay region. He noted, 'An immense number of turtle doves…which fed in part

Passenger Pigeon

Louis Agassiz Fuertes

Great Egret

Photo: Saul Bocian

in acorns.' The Passenger Pigeon was a large bird, much larger than the Mourning Dove. It nested in Southern Ontario including most of Georgian Bay, east to the Maritimes and north to Moose Factory. The seemingly limitless flocks, miles wide and long, were described as a 'biological storm.' Tree branches snapped under the collective weight of countless birds as they descended to feed on the beechnuts, acorns and other hardwood fruits. They nested in huge numbers and native people fed for weeks on the young pigeons, which they dislodged from nests with sticks. Their nests were flimsy structures and they usually laid only one egg. In such large numbers they were easy to shoot and market hunting became a flourishing business. As the forests were cleared and forest fragmentation accrued, the massive populations showed signs of stress.

In 1876, a local newspaper described the wanton shooting of pigeons in Owen Sound. This harvest was occurring throughout the east and coupled with habitat loss of hardwood forests and other factors, the species was in trouble. The last breeding record for Ontario was in Frontenac County in 1898, more than 100 years ago. The last two Passenger Pigeons ever to be seen in Georgian Bay were observed at Penetanguishene on 18 May 1902. It is thought to be the last time that the species was seen in Canada.

Penetanguishene and Georgian Bay hold another ornithological record of dubious distinction. Georgian Bay was not visited by a *'trained and knowledgeable ornithologist'* until about 1825, when Dr. John Richardson, an eminent naturalist, visited Penetanguishene on a leg of the Second Overland Franklin Expedition (1825-1827) collecting birds such as Eastern Bluebird *(Sialis sialis)* and Dark-eyed Junco *(Junco hyemalis)*, the oldest specimens ever recorded for Southern Ontario and Georgian Bay.

Gulls

Gulls and Terns belong to a bird group called the Larids. The Ring-billed Gull and the Herring Gull are the most common gulls in Georgian Bay colonizing the outer islands in large numbers. The gregarious Ring-billed Gull nests on the ground within pecking distance of other birds. The Herring Gull is larger with flesh-colored legs and a prominent red spot on the end of the lower bill while the Ring-billed Gull has a black ring around the tip of its bill. Both have adapted very well to landfill sites and other human development. The largest colony of Herring Gulls in Ontario is on Chantry Island in Lake Huron at Southampton, with more than 3,000 nests and is one of the largest colonies anywhere in Canada. Other large colonies occur in Georgian Bay on The Fishing Islands off the eastern shore of the Bruce Peninsula. Herring Gulls remain in the Great Lakes for the winter but Ring-billed Gulls migrate to warmer areas such as Florida.

Great Egret

One of the most striking birds, the Great Egret, is now a summer resident of southern Georgian Bay, nesting in increasing numbers on Nottawasaga Island near Collingwood. In 1985, there was one nest. In 2005, there were more than 100 nests and by 2009, there were 140

Hummingbird

Photo: Ann Brokelman

nests, making it the largest colony in Canada. For the last ten years Great Egrets at this site have been banded with red color bands and observations were recorded. These egrets have been seen in winter as far away as North Carolina and Cuba. In November 2005, a Great Egret banded in Georgian Bay in June was spotted by Portuguese birdwatchers at Mosterios on the western side of Sao Miguel Island in the central Azores. This is a flight of almost 4,000 km.

Hummingbirds

Sugar water feeders attract hummingbirds to backyards everywhere in Georgian Bay. Red is the most attractive feeder color because the hummingbird selectively sees things in the UV range. There are few red-coloured wildflowers in Georgian Bay, and they have developed this behaviour over countless generations spent on the wintering grounds in places like Mexico where red flowers are common. Georgian Bay is home for only three to four months of the year but

Trumpeter Swan

Photo: Janice Melendez

researchers are now encouraging the extended use of nectar feeders beyond the traditional Mother's Day to Labour Day. Hummingbirds are aggressive to each other and have a unique display when provoked, flying in arcing 'U' shaped flight that they execute with great speed.

The rise of the Trumpeter Swan

Trumpeter Swans are the most notable bird species on Georgian Bay as they are the largest species of waterfowl in North America with a wingspan of up to two metres and weighing up to 12 kg. They are a majestic species and spectacular in flight with their white feathers and contrasting arrow-like black bill. Their most distinguishing feature is their booming voice, which is like a single note from a French horn. A large flock of trumpeters sounds like a barrage of old car horns. Once abundant, Trumpeter Swans have returned to Georgian Bay after an absence of more than 200 years assisted by an innovative re-introduction program. The missionary, Brother Gabriel Sagard, made the first notations of swans occurring in Georgian Bay in the 17th century. Archaeological excavations at Ste. Marie uncovered 178 Trumpeter bones, representing at least 60 birds. It appears that Trumpeter Swans were an important meat source at the 17th century mission, possibly more so than White-tailed Deer. The extirpation of the Trumpeter Swan was almost entirely due to overharvesting, well before significant wetland loss and pollution. The swan was an easy mark after firearms were introduced. They were harvested for their meat, for their feathers used in the millinery business, and for quill pens because Trumpeter Swans have more feathers than any other bird species known, about 25,000. Trumpeter Swans are now found throughout Georgian Bay.

Osprey

There are few more spectacular avian sights than an Osprey making a dramatic dive from up to 40 metres above the bay to pick up a fish in its sharp talons. More than 100 years ago, the Osprey was a common resident of Georgian Bay but by the 1960s it had disappeared as a breeding bird and it was not until 1973 that a single pair nested again. By 1993 there were more than 50 pairs nesting in Georgian Bay. Their extirpation was, for the most part, attributable to the widespread use of pesticides. The Osprey's opposable outer toe can be moved forward or backward and spines on its feet help to secure the prey. The Osprey weighs up to 1 kg and has a wingspan of more than 1.5 metres. Equally spectacular is its annual migration of 8,000 km to and from Georgian Bay as it makes its way to winter in South and Central America and back again each spring and fall. The young will take three to five years before they nest and will usually reside within 50 km of their birth site. Half of young Ospreys die in their first year and about 10-15%

Turkey Vulture

Photo: Janice Melendez

of adults die each year. The average life span is six to ten years. They utilize the shallow inshore area and are quite comfortable near human settlement.

Turkey Vulture

Turkey Vultures are carrion-eating birds and their numbers are increasing steadily in Georgian Bay. They were rarely seen here until the 1950s but the Niagara Escarpment offers many thermals for their soaring flights and nesting opportunities are abundant. They are now a common sight over Georgian Bay from spring to fall and are easily recognized by the dihedral (V) wing pattern that spreads almost 2 metres. The population of these massive birds has increased quickly due to increases in White-tailed Deer, the clearing of forests and the proliferation of roads that has resulted in more road kills. Turkey Vultures nest in old buildings, especially old barns, or cliff edges and caves typical of the Escarpment. They are ugly for a reason; their bald red heads and probe-like beaks are adapted to probing carcasses. The Turkey Vulture soars on thermals, gliding effortlessly, tipping the dihedral from side to side. They have an interesting and sickening defense; they vomit a wad of stinky, partially digested meat. It also urinates on its legs as a means of cooling them as the urine evaporates and it also serves to kill off bacteria picked up from standing in decaying carrion. They have good eyesight and a good sense of smell, the latter being unusual for birds. Turkey Vultures cannot lift or catch prey with their feet and must step on their food to keep it in place.

The Important Bird Area (IBA) program

This is a global initiative to conserve the world's most significant sites for bird populations and there are 13 IBAs on or near Georgian Bay that provide habitat for nesting or migratory birds and are used by threatened species, endemics, and representatives of an area or as staging areas for significant populations. These are:

1. Cabot Head, Upper Bruce Peninsula
2. Owen Channel, Wikwemikong
3. Spring Bay, Manitoulin Island
4. Manitoulin Island North Shore, Gore Bay
5. Lake Huron, Quarry Bay, Meldrum Bay
6. The Cousins and Blind River
7. St. Mary's River Complex, Echo Bay to Maxton Bay, Sault Sainte-Marie
8. Limestone Islands, Snug Harbour
9. Matchedash Bay, Waubaushene
10. Wye Marsh, Midland
11. Nottawasaga Island, Collingwood
12. The Watchers, Penetanguishene
13. Tiny Marsh, Elmvale

The IBA at Cabot Head at the northern tip of the Bruce Peninsula extends from Halfway Log Dump to Cape Chin, and including Dyer Bay is one of the most spectacular parts of the Bruce Peninsula where steep cliffs drop into Georgian Bay. It is the staging for hundreds of Red-necked Grebes, one of the most significant concentrations on the continent, and the Common Loon and the Long-tailed Duck.

The Watchers (north and south) are a group of islands in the southeastern corner of Georgian Bay not far from the entrance to Severn Sound. The South Watcher is a smaller cobblestone shoal while the northernmost is a limestone shelf (alvar) with limited shrubby vegetation. Here there are regionally significant populations of Caspian Tern, Ring-billed Gull and Cormorants, Song Sparrows, Eastern Phoebe and Spotted Sandpipers.

Double-crested Cormorants

The Double-crested Cormorant is one of the most noticeable birds on the islands of Georgian Bay with some colonies containing more than 1,000 nests. But it is an invasive species which first appeared in the early 1930s following an eastward colonization through Lake Superior and Lake Nipigon. They had been nesting in northwestern Ontario in the Lake of the Woods for hundreds of years and their eventual invasion into the Great Lakes increased so much that by the late 1940s control measures were authorized to both appease public concern and reduce suspected competition with commercial and sport fish populations. Population controls took place in Georgian Bay until 1966, using a variety of techniques that had little effect but cormorant numbers began to decline dramatically during the '60s and early '70s. The clue lay in their eggs. Eggshells were about 30% thinner than normal by 1973, causing the eggs to break; cormorants stand on their eggs during incubation, wrapping their webbed feet around them. DDE, a metabolite or breakdown product of DDT was the culprit. With the banning of these substances in 1974, cormorant

Double-crested Cormorant

Photo: Homer Caliwag

Pileated Woodpecker
Photo: Janice Melendez

populations began to rise with numbers doubling every three years. Their meteoric rise was augmented by an increase in small fish such as Rainbow Smelt and the Alewife, their primary food source occurring in large schools in shallow water.

The Double-crested Cormorant, once rare in Georgian Bay, is now more abundant than ever.

Woodpeckers

Woodpeckers are important habitat enhancers as their nest sites are used by other animals. The Northern Flicker excavates a cavity in Basswood trees used by arboreal bats such as the Silver-haired Bat. There are nine species of Woodpecker found in Georgian Bay. The Downy is the smallest and the Pileated is the largest. These two species are present all year in Georgian Bay. The Yellow-bellied Sapsucker is migratory and arrives back in April.

Prairie Warbler

The Prairie Warbler is not a resident of the Prairies but is at home in parts of the rocky shores of the 30,000 Islands. The largest population of Prairie Warblers is found on the eastern shore of Georgian Bay, one of only a few areas where this species breeds in Ontario. Amid rocky outcrops, it nests in stunted White Pine, Red and White Oak and juniper bushes. The largest population can be found in the Cognashene-Honey Harbour area as well as in Massasauga Provincial Park, Go Home Bay and Sans Souci. After a summer on Georgian Bay, it may winter in the Bahamas, on Caribbean islands, and in southern Florida. Before European settlement, the species was rare or absent in much of its present breeding range but by the mid-twentieth century, following deforestation, it became widespread.

Shorebirds

The three main groups of shorebirds are the curlews, the plovers and the sandpipers. Some shorebirds are adapted to upland living. Killdeer, Spotted Sandpiper, Upland Sandpiper, Piping Plover, American Woodcock and Wilson's Snipe all stay in the area. The small shorebirds that may pass through in large flocks, include the Least Sandpiper, Semipalmated Sandpiper, White-rumped Sandpiper and Sanderling. Shorebirds migrate long distances and for most species, the adults leave about one month before the young birds, heading

Prairie Warbler

Photo: Randy Holland

south in late June. The last shorebird to be seen in Georgian Bay is the Purple Sandpiper which leaves in October and November.

Great Blue Heron

This is one of the largest colonially nesting waterbirds in Georgian Bay feeding on small fish, amphibian, insects, crayfish and snakes. A decline in amphibian species such as the Green Frog may be the cause of a recent decline in numbers.

Terns

Terns are gull-like birds that are more graceful and more slender than gulls. With their sharp pointed bills, long pointed wings and often deeply forked tails they are often referred to as 'sea swallows.' The Caspian Tern is the largest tern in the world and is a typical island dweller often feeding in waters more than 40 km away from its nest. Those on the Watcher Islands will feed in Matchedash Bay near Waubaushene. As with all colonially nesting waterbirds, their population can rise and fall quite quickly.

Swallows

Swallows are adept at catching insects and they are much more efficient than bug zappers. Unfortunately, they are becoming less abundant, as are all of the birds called 'aerial foragers', who 'hawk' for insects on the wing. Barn Swallows were probably not very abundant prior to European settlement in Ontario and are now found in barns, boathouses and other man-made structures. Bank Swallows nest in sandbanks and gravel pits as their name implies while Cliff Swallows favour the underside of bridges and are usually not found near cliffs. Their mud nest is wider at the bottom with an opening that may jut out to the side or downward. The Northern Rough-winged Swallow nests in sandbanks, but unlike the Bank Swallow, it does not nest in colonies, preferring to nest singly and often choosing an old cavity built by the Belted Kingfisher. The Bank Swallow is the smallest of the swallows while the Purple Martin is the largest but has declined significantly and its northern range in Ontario has shifted south.

The Tree Swallow, with its iridescent blue-green back and white undersides, is the most abundant of the Georgian Bay swallows and it will nest around homes and cottages. It is attracted to nest boxes.

Gulls on the Watcher Islands

Waterfowl

The waterfowl of Georgian Bay include ducks, geese and swans of which the most familiar are the dabblers, those species so-called because they rarely go completely underwater but feed on the surface and just beneath by bobbing upside down, such as the Mallard. The most common species of waterfowl in Georgian Bay are the mergansers and the Wood Duck which is the most coloruful. Several species of waterfowl nest in tree cavities. The Hooded Merganser is perhaps the most versatile of all waterfowl species in Georgian Bay as it frequents all major habitats: open water on migration, the forested edges of waterways, shorelines and wetlands for raising young. But while it is quite common in Georgian Bay, it is secretive and is not often observed.

Hooded Merganser

Photo: Janice Melendez

CHAPTER

7

THE PEOPLING OF GEORGIAN BAY DURING THE LAST ICE AGE

PETER L. STORCK

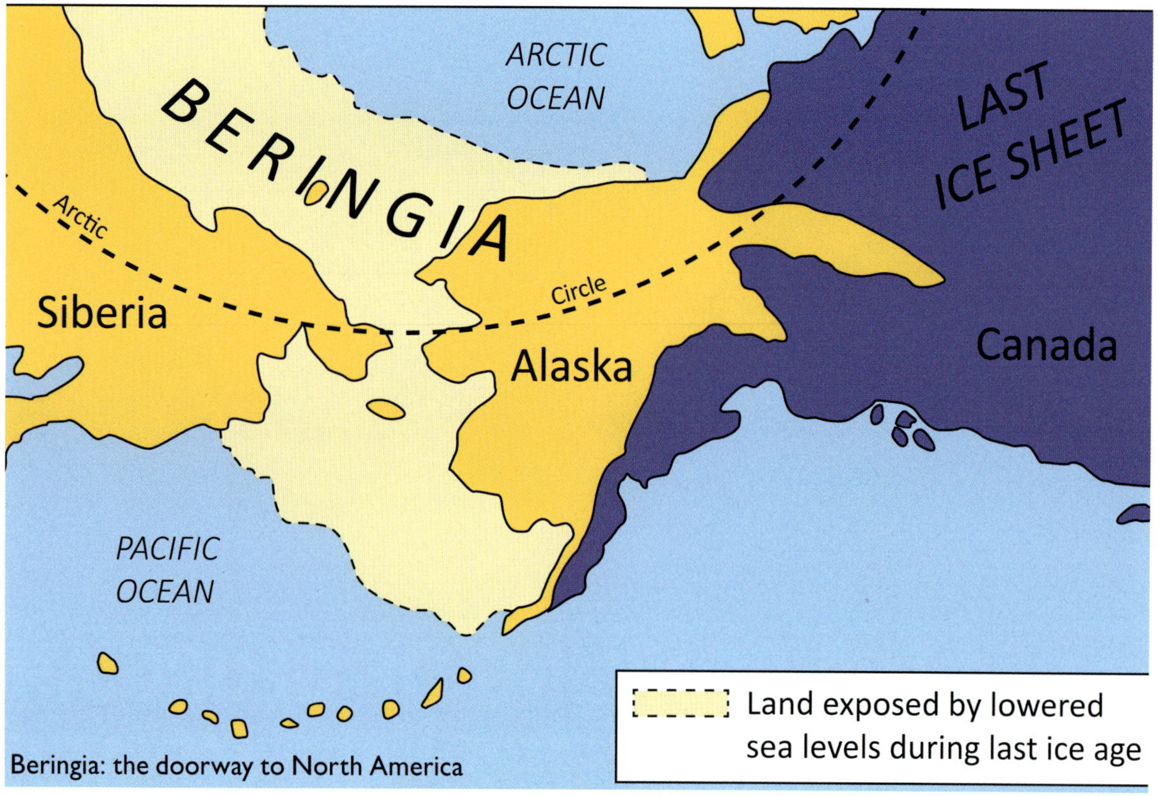

Beringia: the doorway to North America

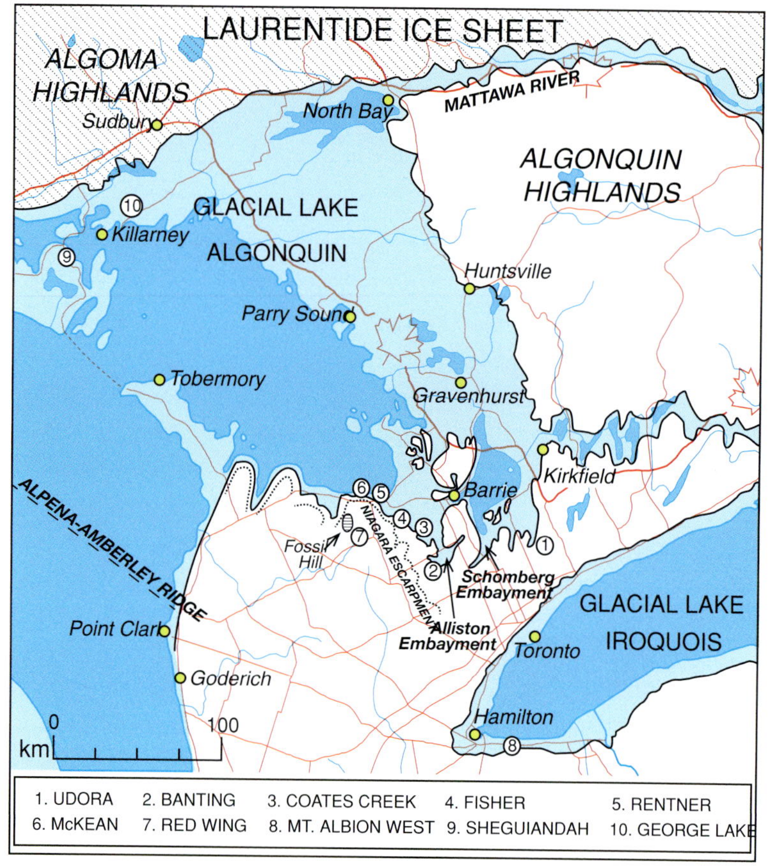

Ontario at the end of the last ice age: a land of lakes

The Ancient Ones.

They gathered at the fire and cast in the feet of the animals they had killed.

They did this out of respect so the flames would carry the spirits of the animals from this life to the next life and their feet could carry them back to this life back to the people, their children and their children's children.

In this way life would continue as it had from the beginning when the Great Spirit created the world and the spirits of all living creatures.

This story, though fictional, reflects widespread beliefs among historically-known hunter-gathers that may also have been held by Paleo-Indians.

Colonizers of an Ice Age Landscape

When ice age hunter-gatherers of northeast Asian descent first appeared on a small peninsula of recently deglaciated land that we know as Southern Ontario, all of the remainder of Canada (except for part of the Yukon) was deeply buried under a continental ice sheet. These ice age hunter-gatherers were first discovered archaeologically in the 1930s in the American Southwest where very distinctive and well-crafted spear points were found in kill sites associated with the bones of mammoth and extinct forms of bison. The people who made these spear points and other tools were called Early Paleo-Indians by archaeologists and, in later decades, were discovered to have occupied almost all of North America between 11,500 and 10,400 years ago extending from Alaska to Central America. In that entire region they were one of the most important founding populations for Native Americans, possibly in South America as well. It is a lasting irony that archaeologists have not as yet identified the ancestors of Early Paleo-Indians; though they had Asian origins there are no Early

Paleo-Indian sites in northeast Siberia and the culture appears to have evolved in North America.

When the first Early Paleo-Indians arrived in Southern Ontario they encountered the water world of glacial Lake Algonquin. The lake, bordered in northern Ontario by the retreating ice sheet, occupied the Huron and Georgian Bay basins, encompassing the Simcoe lowlands as well, and extended as long embayments named after the modern communities of Alliston and Schomberg – almost to the northern margins of the future Greater Toronto Area.

Lake Algonquin began as an ice-marginal lake in the southern part of the Huron basin around 12,000 years ago. As the ice retreated northward it inundated much of the isostatically-depressed (and low-lying) Bruce Peninsula and expanded into the Georgian Bay basin and surrounding lowlands, with an outlet in the vicinity of modern-day Kirkfield on the Trent-Severn Waterway. From there, Lake Algonquin water drained down the Trent River and into Lake Iroquois in the Ontario basin, with an outlet to the Mohawk River Valley in New York State and then down the Hudson River Valley to the Atlantic Ocean. The Kirkfield outlet closed around 10,400 years ago, possibly due to uplift of the land following the release of the weight of the ice sheet, at which time Lake Algonquin found new and lower outlets in the North Bay area and drained down the Mattawa and Ottawa river valleys to the St. Lawrence. It was probably during this stage of Lake Algonquin that Early Paleo-Indians in the mid-continent region to the south entered the province. And it was because Early Paleo-Indians camped on the shores of this glacial lake that several important archaeological discoveries were made in the 1970s and 1980s, giving us revealing glimpses into Early Paleo-Indian life.

Life on a Glacial Lake

One of these discoveries is the Udora site, named after a nearby community and located a few kilometres southeast of modern Lake Simcoe. The archaeological site consists of eleven more or less distinct concentrations of broken tools and stone debris from tool making that occur along a former shoreline of Lake Algonquin. One area produced a large cache of tools; actually, simply spalls from larger blocks that were used as tools without any shaping. Microscopic analysis of the damage on the edges of the tools indicates that they were used by two individuals of different strengths to split spruce roots, implying different ages (possibly two males or two females working together, and kin related). Ethnographically, in hunter-gatherer societies, this activity was most often done in spring when the roots were most pliable, and by women to produce binding or weaving material. The tools were buried together but in separate clusters, thus probably in skin bags, for re-use when the individuals returned to the site. Apparently they never did, leaving an untold story.

Excavation also exposed areas where spears were repaired and animal hides may have been cleaned for use in clothing, bedding, or tents. A fire pit revealed a number of tool fragments and debris from tool making as well as fragments of spear points broken in the hunt and returned to the camp still in their hafts for replacement. The fire pit also produced highly fragmented and calcined (burnt) bone–the first obtained from an Early Paleo-Indian campsite in the Great Lakes region. Among the bones were fragments from the foot of a caribou (*Rangifer tarandus*), the first direct evidence that Early Paleo-Indians hunted caribou in the Great Lakes region.

Other bone fragments are from a hare (either snowshoe or

Excavation of the Udora site *by the Royal Ontario Museum, 1987. Courtesy of the Royal Ontario Museum.*

Right: **Telltale bits and pieces** recovered from a Paleo-Indian site. Courtesy of the Royal Ontario Museum.

Above: **Early Paleo-Indian caribou hunters.** Courtesy of the Royal Ontario Museum. Illustration of Early Paleo-Indian caribou hunters by Ivan Kocsis.

arctic) and, most unexpectedly, a foot bone from an arctic fox (*Alopex lagopus*). The hare and fox are typical of northern tundra today and may have been trapped for their furs. Eleven thousand years ago, the margin of the ice sheet was roughly 200 kilometres north of the site, running in an east-west direction across the top of the future Georgian Bay and Algonquin Park. Cold air flowing off the ice sheet may well have created tundra conditions far south of the ice mass.

Another site associated with glacial Lake Algonquin, but closer to the margin of the future Georgian Bay, is the Fisher site named after the owner of the property, located 12 kilometres south of Collingwood. It was occupied by the immediate descendants of the people who occupied Udora judging from the spear points which, although technologically similar, were smaller and differed in other details from those made by their ancestors. Some archaeologists think this change reflects a transition from throwing or thrusting spears to the use of a throwing stick (atlatl) to propel darts.

The Fisher site is among the largest Early Paleo-Indian sites in North America. It consists of tool fragments and the debris of tool making scattered over at various sites across an area of roughly 22 hectares. Excavations indicate they were used for different activities: some for the early stages of tool manufacture, others for tool repair, hide preparation, and probably a range of domestic activities. Microscopic analysis of use-wear damage indicates that some tools were used in filleting fish.

These activity areas seem to occur in pairs, as though complimentary activities, representing short-term occupation by separate family groups, possibly at one time as part of a band of allied families, or they could represent separate occupations by individual families or task groups over a long period of time, perhaps even different seasons. This kind of ambiguity is not uncommon in archaeological interpretation, limited as it is by post-depositional disturbances that confuse the archaeological record and the inability of radiocarbon dating (assuming organic preservation, a rarity) to discriminate between events less than several decades apart.

The Fisher site was situated on the margin of a lagoon impounded by sand bars on the shoreline of glacial Lake Algonquin. Fossil pollen suggests that the lagoon was surrounded by a spruce parkland, with open expanses of sedges and grasses on hilltops and exposed slopes and copses or groves of spruce in low-lying protected areas. This type of habitat is also consistent with the arctic fox found at Udora, a species that today lives in the tundra but during late winter or spring may occupy tree line/forest edge habitats.

Some tools were used in filleting fish and this may have been the primary reason the site was occupied. The highest portions of the site may also have served as game lookouts for monitoring the movements of any caribou as they foraged along the Algonquin strandline.

The southernmost margin of glacial Lake Algonquin had two embayments: the Alliston embayment (named after the modern town) and a short distance to the east, the Schomberg embayment, now Holland Marsh. In the Alliston embayment, a small Early Paleo-Indian site – Banting, named after the landowner, Edward Banting,

***The Early Paleo-Indian tool kit** at the Fisher site consisting of (upper left to right) tools for slotting, piercing, scraping and (lower) hammering, projectile points and dressing hides. Courtesy of the Royal Ontario Museum.*

the nephew of Sir Frederick Banting, the co-discoverer of insulin – occurs on a drumlin which formed an island in the lake. The site may have been occupied in winter as a lookout to monitor caribou movements on the ice, a behaviour caribou adopt to avoid wolves. To the east, in the Schomberg embayment, several Early Paleo-Indian sites are associated with the Algonquin strandline near Schomberg. One archaeologist, citing hunting strategies used by modern arctic caribou hunters, has suggested that the clustering of sites in the area may represent spring occupations by Early Paleo-Indians attempting to intercept the northward moving caribou during seasonal migration. This interpretation demonstrates the value of large numbers of archaeological sites for revealing patterns of land use and raising new questions for research.

***Campsite activities** at the Fisher site. Courtesy of the Royal Ontario Museum. Illustration of Campsite activities at the Fisher site by Ivan Kocsis*

For archaeologists, using the Lake Algonquin strandline as a way to enter the Early Paleo-Indian world was successful, but these hunter-gathers almost certainly used other parts of the landscape. The problem for archaeologists is knowing where to look. The solution to filling in this knowledge gap was both surprising and richly informative.

From the Beaches to the Uplands: Mining comes to Georgian Bay

Hunter-gatherer technology is heavily dependent on stone tools made from flint or a similar material, chert. Early Paleo-Indians in Southern Ontario preferred a white-coloured chert which dominates their tool kits at sites in both central and southwestern Ontario. Years of geological field work succeeded in identifying the only source of this material: the Fossil Hill Formation near the top of the Niagara Escarpment, south of Thornbury.

The Fossil Hill Formation is exposed along the face of the Niagara Escarpment and contains chert nodules and lenses composed of glass-like silica. It is thought that the original source of the silica was the Precambrian Lorrain Formation quartzites of the La Cloche Ranges in northern Georgian Bay which emerged as islands in the warm Silurian seas that once covered Georgian Bay some 420 million years ago when Ontario straddled the equator. Weathering in a warm tropical climate allowed silica-rich waters to runoff to the sea. This would have triggered blooms of silica-secreting micro-organisms in the water column which died and settled to the sea floor. These were dissolved later at shallow depths creating silica-rich waters which moved through the rocks to slowly build up nodules and lenses of silica.

The discovery of what are the oldest Fossil Hill quarries in Ontario along the Niagara Escarpment led to the finding of other Early Paleo-Indian sites: notably Red Wing named after a nearby village and located below a chert outcrop, which was clearly occupied for the purpose of obtaining toolstone and working it into tool blanks (for transport elsewhere), or directly into tools to replenish the toolkit. Exhausted tools were discarded on the site. One of the most surprising archaeological discoveries was several fragments of chipped-stone wood-working adzes, probably made and used only at Red Wing since tools of this type have not been found on Early Paleo-Indian sites elsewhere in Southern Ontario.

Red Wing poses many questions that the excavations were not able to answer. Chief among these is how Early Paleo-Indians discovered the chert source and why they preferred this material over another type of chert that occurs along the north shore of modern Lake Erie. Both types are equally workable although it's possible the Lake Erie sources were under water and not accessible 11,000 to 10,400 years ago. The Fossil Hill chert in the Georgian Bay region may have been more closely related to their annual subsistence cycle. The chipped stone adzes suggest that the uplands had other resources of value, perhaps wood with narrow and densely compacted growth rings from trees that grew slowly in exposed habitats. Being harder than wood from trees growing in more protected environments, it may have been especially valued for spear and dart shafts, even handles.

Winter Refuge

Early Paleo-Indians did not visit the Red Wing site in winter months when the rocks would have been deeply buried under snow. Today, the area is a prominent snowbelt because of snowfall on the top of the Niagara Escarpment from moisture-laden winds blowing off Lake Huron. This snowbelt was undoubtedly more extensive during the late ice age so the question is, where did Early Paleo-Indians live in winter? The answer might be found along the western and northern shores of early Lake Ontario, roughly 140 straight-line kilometres south of Redwing. This was suggested by discoveries at the site of Mount Albion West at the western end of the Niagara Peninsula, near Hamilton.

The Mount Albion West site occurs near the crest of the Niagara Escarpment. It is adjacent to Red Hill Creek that descends into the basin of modern Lake Ontario in a steep-sided and gradually broadening valley. This strategic location may have been selected by hunters who intended to ambush animals, possibly caribou or even mastodon or mammoth moving between the top of the escarpment and the lake below. Sitting on an abundant supply of stone useful for making tools, the site may also have given people an opportunity to replenish their tool kits before descending into the lake basin for extended hunting or even over-wintering. At the time, lake levels in the Ontario basin were very much lower than today, exposing much of the lake floor that may have provided good grazing and browsing habitats, not to mention a more protected environment. Unfortunately, this is little more than informed speculation because no animal bone was preserved at the site and no evidence was found (such as toolstone from bedrock formations exposed deeper in the lake basin, now covered by water) that would confirm that the people at Mount Albion West actually descended into the basin. The possibility raises questions for future research should other sites in comparable geographic locations ever be found – by archeologists, cottagers, or construction workers.

Top: ***Technological change*** *recorded in stone; the shift from unnotched to notched bases on spear or dart points. Courtesy of the Royal Ontario Museum.*

Later Peoples and the Post-Algonquin Lakes

As glacial ice retreated and lower outlets for Lake Algonquin were uncovered in the region of North Bay, lake levels began receding, producing a number of early Holocene lakes that, in their descent, shrank progressively closer to the modern-day margin of Georgian Bay. Many of those early lakes were not only occupied by Late Paleo-Indians, but by later Early Archaic hunter-gatherers also adapted to the changing environment.

One site from this time period in the southern part of the Georgian Bay region is Coates Creek, about 20 kilometres south of Collingwood. The site occurs on the south end of a peninsula that jutted out into a post-Algonquin Lake – possibly Lake Ardtrea.

Late Paleo-Indian spear points and knives or tool blanks from the Sheguiandah site, Manitoulin Island. Courtesy of the Royal Ontario Museum.

Spear points show a mixture of Late Paleo-Indian and Early Archaic technology reflecting a culture in transition.

Another site in the same region, McKean, also occupied by Late Paleo-Indian/Early Archaic peoples, is a short distance east of Collingwood and roughly a kilometre inland of the modern lakeshore. Although the site is near a strandline that may have been formed between 6,000 and 4,000 years ago, the occupation clearly pre-dates that time period; animal bone are from fish as well as deer and either moose or elk. Thus, unlike the Early Paleo-Indians before them who hunted caribou and trapped arctic fox in a parkland habitat near a tundra that fringed the ice sheet, glacial retreat and environmental change forced Late Paleo-Indians and Early Archaic people to exploit an essentially modern fauna, presumably requiring a change in their hunting techniques as well.

South of Coates Creek and McKean, Late Paleo-Indian and Early Archaic hunter-gatherers also re-occupied a number of Early Paleo-Indian sites in the Alliston and Schomberg embayments. During post-Algonquin times, these sites probably occurred in a closed forest and the people used those locations, and former strandlines, to hunt modern fauna in marsh environments.

Northern Georgian Bay: A Record of Rapid Environmental Change and Adaptation

Northern Georgian Bay was only occupied by humans after glacial Lake Algonquin had dropped to lower levels, creating a succession of brief postglacial lakes. The first archaeological evidence for early humans in the area was discovered in the 1940s at the George Lake 1 and 2 sites located on the slopes of Killarney Ridge in the La Cloche Mountains of Killarney Provincial Park. A decade later, the Sheguiandah archaeological site was discovered on the eastern end of Manitoulin Island. All three sites were probably occupied, at least in part, because they occurred on sources of stone useful for making tools, in this instance, quartzite. The George Lake sites, with their assemblages of stone tool fragments and knapping debris, are silent witnesses to the human presence. But Sheguiandah, with its provocative stratigraphy and paleoecological record, opens a window into one of the most dramatic periods in the Great Lakes human experience, marked by large fluctuations in lake levels and rapid environmental change.

Initial excavation of Sheguiandah in the 1950s suggested broken tools and debris of stone tool making occurred in till, glacial sediment deposited during the last ice age, and thus that human occupation pre-dated the last glaciation. Since this advance was thought to have occurred at least 30,000 years ago, the human occupation was that age or older. This would have made Sheguiandah roughly three times older than any other known archaeological site in the Americas and for this reason the excavation became highly controversial.

Re-excavation of the site four decades later showed that tool fragments and debris are much younger since they are found not in till but in postglacial lake sediments, deposited in a wave-churned near-shore environment (hence the mixed and jumbled nature of the sediments): they are around 9,700 years old. At that time, the newly emerged landscape was in transition from open spruce parkland (similar in appearance to the forest-tundra transition today) to a pine-dominated forest. Broken spear points discarded at the site suggest that it was occupied by Late Paleo-Indians, very possibly having just colonized the region from the Plains (and thus also referred to as Plano by archaeologists). Because of their association with the early Great Lakes, these people have also been called Aqua-Plano in this region.

The Missing Millennia

The history of Lake Huron and Georgian Bay between about 10,000 – 4,500 years ago is one of marked changes in climate recorded by abrupt changes in lake level. A cool and dry climate between 7,800 and 7,500 years ago followed by warm and dry conditions after 7,500 before present, resulted in at least three low-stands in the Huron/Georgian Bay basins called Lake Stanley in the Huron basin and Lake Hough in the Georgian Bay basin – both named after pioneering geologists. Low lake levels were also aided by the continued depression of the North Bay-Ottawa region in response to the great weight of the ice sheet though this area was recovering (rebounding) quickly.

Low-stands when Lake Huron was as much as 30 metres lower than today occurred between 9.8-9.5 thousand years ago (Early Stanley), 9.4-8.1 (Mid Stanley), and 8.0-7.5 (Late Stanley) such that Lake Huron received no discharge from Lake Superior and did not discharge downstream; Erie would have been a hydrologically-closed basin as well and Niagara Falls would have been dry. The climate during these periods was as much as 35% drier than today and the lakes, experiencing higher evaporation than precipitation, were brackish. It would have been possible to walk across the then submerged Niagara Escarpment from Manitoulin Island to the Bruce Peninsula.

In the central part of Huron basin, now flooded by the waters of Lake Huron, side-scan sonar and underwater photography suggest that Late Paleo-Indian/Early Archaic people may have built hunting blinds and game drives on that now submerged corridor (today, at least 30 metres underwater), ostensibly to hunt caribou, although no remains have been found.

Dramatic swings in lake level between 10,000 and 4,500 years before present mean that several thousand years of human prehistory around the Georgian Bay basin is now either underwater or deeply buried beneath lake sediments, or has been washed away. First Nation Anishinaabeg oral traditions – stories passed from one generation to the next about their creation and relationships with the world they

Left: ***Quartzite point*** *from the village of Flesherton at the base of the Bruce Penninsula. Courtesy of the South Grey Museum.*

Right: ***Extraordinary Early Paleo-Indian projectile point*** *made of quartz crystal. Areas of dark brown and bright-reddish brown colour may be traces of hematite from blood used in a hafting cement, and red ochre used as a pigment for ceremonial purposes. Courtesy of Arthur Roberts and Royal Ontario Museum.*

inhabit – hint of fluctuating water levels and a passage of land used by their ancestors for walking between Manitoulin Island and the Bruce Peninsula. Whether or not these oral traditions can be regarded as reflecting historical experiences, they certainly reveal an inseparable cultural and spiritual bond between First Nations and the land, a bond they depend on – as they always have – for their very survival as a people.

The Story Continues: Prehistoric Peoples in Modern Georgian Bay

Over the last several thousand years, during the period when Georgian Bay and the Great Lakes attained their present configurations, regional climate underwent a cooling trend producing hardwood forests around the southern portion of Georgian Bay and boreal forest with an intermediate zone of mixed hardwoods and conifers to the north. During this period descendants of the ice age colonizing peoples adapted to broad ecological zones: Shield Archaic people foraging across the boreal and mixed hardwood-conifer forests and people of the Laurentian Archaic living among the deciduous forests to the south. Georgian Bay was occupied by both social groups, evidence of Shield Archaic people occurring in Killarney Park and Laurentian Archaic on the Bruce Peninsula and the south shore.

Very little is known about the lifeways of these people other than the fact they were hunter-gatherers and, seasonal fishers as well, an activity that probably had its beginnings during the Paleo-Indian occupation of the region. Indeed, fishing may have become increasingly important over time as the Great Lakes were re-populated by fish from ice age refugia in the Mississippi drainage and the Atlantic. Roughly 6,000 years ago, at the Rentner site east of Collingwood, middle Archaic people obtained whitefish in the fall and, in spring and summer, catfish, sucker, and drum (silver/gray bass). They also hunted deer and trapped porcupine, beaver, and hare. There is evidence elsewhere for an increasing amount of social interaction in an expanding population across the mid-continent region, creating a complex social landscape. This would have profound consequences for future cultural and political developments in prehistoric North America and most would be echoed in the Georgian Bay region.

Painting of creation of Middle Woodland mounds

CHAPTER

8

FARMERS, FISHERS, HUNTERS AND TRADERS: INDIGENOUS COMMUNITIES ON GEORGIAN BAY

RON WILLIAMSON
MARTIN COOPER

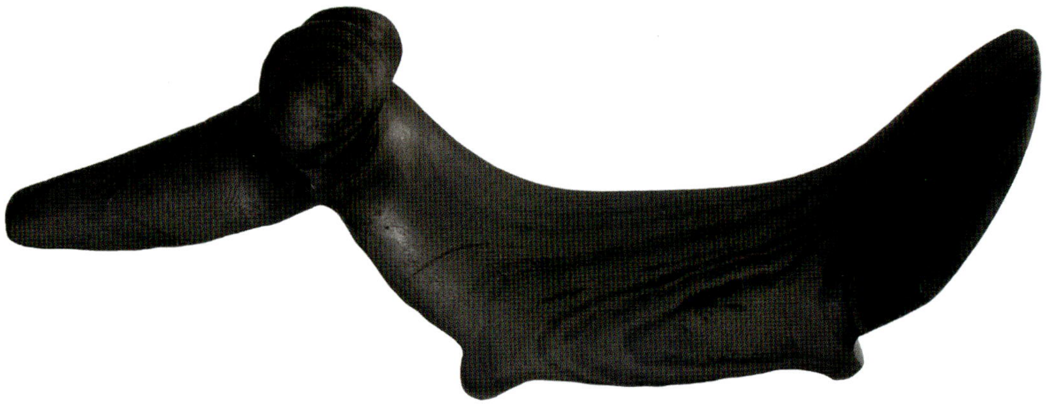

Left: **Birdstone,** *Wilfrid Jury Collection.*

Right: **Meadowood cache,** *Peace Bridge site.*

Cultural Change Comes to Georgian Bay

About 2,500 years ago, profound technological and social changes were taking place across the southern margin of the Canadian Shield. Ceramic vessels used for cooking and to store water were introduced into the Great Lakes region from the southeast United States, signalling the growing importance of cooked wild plants in diets. Thereafter, fragments of broken pots became as common on archaeological sites as are the broken stone tools and small flakes created and discarded during flint knapping of silica nodules. This period is also known across the Great Lakes region for heightened interaction with far distant groups, especially in shared burial ceremonies.

Ceramic vessels were pinched and coiled from local clays to form the pots and then smoothed and decorated on their interior and exterior surfaces, usually near their rims. By 700 years ago, the pots were globular shaped and were decorated with incised horizontal, oblique, opposed, and other complex impressions, sometimes reflecting design sequences from far distant communities. Archaeologists examine the variation in design motifs and techniques used to decorate pots to identify the ethnic affiliations of their makers and to date the sites on which they are found.

The seasonal subsistence cycle of this period involved families harvesting wild plants and animals from locations where they concentrate such as wild rice beds or fish spawning runs. Alternatively, small interior fall and winter hunting camps were occupied by single families. Hunting technology was similar to that of previous times with the use of the atlatl or throwing board. A chert arrowhead-tipped dart was used on a throwing platform, which propelled the dart with considerable velocity into the prey. Carved and polished banded slate birdstones and bannerstones were used as atlatl handles, bases or counterweights and are some of the most beautiful carved stone art ever found in northeastern North America. Fishing was also a focus of the annual round and groups employed a wide range of equipment including nets, weighted with notched stones, lines with bone and copper hooks and gorges, and spears with barbed harpoons made of bone.

Georgian Bay in the Great Lakes Social Landscape

Gravesites reveal much about regional interaction at the time. Some grave goods were made of exotic materials that were exchanged over long distances, such as native copper from the Lake Superior basin and silver from near Cobalt, Ontario, both of which were fashioned into ornaments. Galena (lead ore) from the St. Lawrence Valley has also been found. Very thin and finely fashioned bifaces, made of Onondaga chert quarried from around the east end of Lake Erie, were common. Bifaces represent a middle stage in the reduction of blocks of chert into finished tools. Between about 2,700 and 2,300 years ago, people across what are now New York State, Southern Ontario, and Quebec produced extraordinarily thin and finely made bifaces that could easily be used as points or knives or further flaked quickly into scrapers or drills (such as those at – Meadowood Cache – Peace Bridge). Other goods that were placed with the deceased include polished gorgets and birdstones indicating the latter were not only used on hunting equipment but were also highly valued in the afterworld. The contributions of these offerings to the graves of fellow community members suggest that people were mourning the deaths of those other than their immediate family members. It was also common for people to sprinkle ochre (hematite iron) around and on the remains of their deceased as well as their grave offerings. This substance is bright red and likely symbolized blood, restoration, and rebirth.

By 2,500 years ago, these increasingly complex burial ceremonies sometimes involved the construction of burial mounds based on ceremonies originating in Ohio and even the Mississippi Valley. While the use of prominent natural features such as knolls or sand banks for burial had been common practice previously, people began building small oval or round mounds, which averaged ten metres in length and width and rarely exceeded two metres in height. Most were constructed along the shores of Rice Lake (such as Serpent Mounds), the lower Trent River system and the adjacent waterways draining into the eastern end of Lake Ontario. The practice of burying community members in mounds ceased about 1,300 years ago. Some of the grave goods in this period include gorgets of marine shell, elaborately shaped objects made of copper and mica and exotic cherts such as Flint Ridge chalcedony from the Ohio Valley referred to overall as Hopewell by archaeologists. These types of sites are known as Laurel on the Canadian Shield, Saugeen on the Bruce Peninsula and southward, and in eastern Ontario, Point Peninsula.

One of the most interesting sites of this period is situated along the north shore of Georgian Bay on a beach in Killarney Bay. First examined in the 1870s by the Geological Survey of Canada, it was excavated during the mid-twentieth century by the University of Michigan, who believed that burial mounds above the beach were human-made. Subsequent work at the site, by the National Museum of Civilization (now National Museum of History), led researchers to believe that the mounds were natural sand knolls. More recent examinations of the mounds by scholars from Laurentian University, however, yielded geo-archaeological data that indicated sediments had been added to beach ridges to enhance the mounding.

Cache of Middle Woodland bifaces, Port Colborne site

Ceramic vessels recovered at the site belonged to the Laurel/Point Peninsula Tradition, while others were more similar to the Hopewell Tradition of the Ohio Valley. Grave goods included native copper celts, a blocked end tubular stone pipe, and a cache of 67 bifaces. Despite the antiquity of the site, portions of woven baskets or mats and spear points with portions of wooden shafts still attached were recovered. One of the fascinating aspects of this site was the caching of bifaces, which was a common activity between 4,000 and 1,500 years ago (Port Colborne bag cache). Cached bifaces represent a sophisticated ceremonial practice suggesting their manufacture by a limited number of knappers and thereafter distributed across the

Donaldson copper panpipe.

landscape; few if any of the caches contain any of the thousands of small flakes that would have been discarded during their production. Given the hundreds of caches that have been documented, production was occurring well beyond the individual needs of the knapper or the band of which he was a member.

With time there was a change away from caches of bifaces made of local cherts to ones made of exotic flints from Ohio. This would appear to be coincident with the period of substantial cultural influence originating in the west in the upper Mississippi Valley, which was also expressed with mound-building. It is important to note that among both Iroquoian- and Algonquian-speaking societies, the origin of chert or flint is linked with creation stories and the actual spilled blood and broken bodies of important figures in those stories. Perhaps the knappers or hunters occasionally returned to the earth the stone that was so essential in their lives, a giving back of an important cosmological figure's blood or body.

The best known site of this period, the Donaldson site, is just south of Georgian Bay on the Lake Huron shore on the Saugeen River near Southampton which was occupied by a local band during the spring and early summer where riverine resources, most notably spawning fish, were exploited in large numbers. Like other populations living around Georgian Bay, the Donaldson people participated in burial ceremonies that included the provision of exotic grave goods including stone ear spools and copper panpipes.

It is likely that the Algonquian-speaking Odawa, described in seventeenth-century European accounts as living in the Blue Mountain area and along the Bruce Peninsula, were derived from the populations at Donaldson and similar aged sites in the Georgian Bay region. The cherts and other source stone from Donaldson and other regional sites also derive primarily from the Michigan Peninsula-Lake Huron basin area; artifacts have also been found that were made of catlinite, an argillite that turns a beautiful deep rust-red colour when polished, from sources in Minnesota and Wisconsin.

Another population dating to the same time period was situated on the Nottawasaga River in Simcoe County. The Blueberry site, situated in Wasaga Beach, for example, yielded evidence of two hearths along with fragments from at least six ceramic vessels. The design attributes on these vessels can be related to both Saugeen and eastern Point Peninsula styles. Stone tools were also recovered that included thirteen scrapers, drills, and bifaces along with the waste flakes generated by knapping of cherts from the Fossil Hill formation and Kettle Point on the Lake Huron shoreline near Grand Bend. A small sample of calcined animal bone was also recovered from which only white-tailed deer could be identified. The site also yielded one rolled round bead made of native copper. A number of burials have also been documented in the sandy dunes along the beach and lower reaches of the river with grave goods that include antler hafted beaver incisors and projectile points, tools made of elk leg bones, and ground stone celts. All of these sites are thought to be about 2,000 years old.

In the southeast part of Georgian Bay at the mouth of the Severn River, four more sites have been documented, the largest of which is the Baxter site occupied between 2,000 and 1,800 years ago as a late summer/fall camp. The site yielded caches of eight banded slate gorget preforms and thirteen pointed whetstones, indicating that these items were being produced at the site for exchange through wide reaching trade networks. Other trade materials present at the site included native copper from Lake Superior and Flint Ridge chalcedony from the Ohio Valley.

Growing Rather than Gathering Food

While there is evidence of maize being used more than 2,000 years ago in the Great Lakes region based on the identification of maize-based phytoliths or silica-based compounds on the interior surfaces of ceramic vessels, the earliest sites on which carbonized maize kernels or cob fragments were found date to circa 1,400 years ago. For the preceding thousand years or so, maize was a supplement to the diet rather than a mainstay. Isotopic analyses of bone collagen and carbonate of human bone samples from a wide range of sites, undertaken with the permission of First Nations, indicate that it was not until at least 1,000 years ago that maize became the focus of their economies and diet. Once it was regularly grown and harvested close to base settlements, it could be stored for winter use thereby reducing the need to split into small family groups and disperse into the interior. The first base settlements contained a small number of elliptically shaped houses, most measuring approximately six metres in length and three metres across, encircled by flimsy fences or single-row palisades that likely functioned more as windbreaks. The presence of large, deep pits, probably for storing crops, suggests inhabitants committed to agriculture. With time, the adoption of a farming life led people to a significant change in how they traced their descent from their father's to their mother's side. This perhaps developed because of separate activities for men and women whereby women remained in or nearby base settlements most of the year to plant, tend, and harvest crops while men were frequently away to hunt, fish, and raid their enemies. This in turn led men to reside with their wives' families and also resulted in the development of larger house structures and eventually the longhouse, a communal residence in which a grandmother and her daughters' families resided.

All of these changes coincide with the first appearance of the Iroquoian language in the Great Lakes region. Before the introduction of maize, all of the people in north of the Lake Ontario basin were likely Algonquian speakers. At the same time the atlatl was replaced by the bow and arrow for much more effective hunting of large game. Once this new lifestyle had been introduced, the people who were residing along the central north shore of Lake Ontario evolved into the ancestors of the Huron-Wendat and Tionontaté (Petun), while those to their east along the St. Lawrence Valley were the Iroquoian groups first encountered by Jacques Cartier near modern-day Montreal. South of Lake Ontario, in what is now central New York State, populations also adopted this lifestyle and evolved into the ancestors of the Five Nation Iroquois (Seneca, Cayuga, Onondaga, Oneida, and Mohawk) who inhabited Southern Ontario in the mid-to-late seventeenth century.

Maize horticulture is best suited to the arable soils and relatively milder climate found along the southern littoral of Georgian Bay, in contrast to the Canadian Shield on the east and north shorelines and the limestone plains of the Bruce Peminsula and Manitoulin Island. This dichotomy resulted in a symbiosis between Georgian Bay communities and agriculturalists developing along the north shore of Lake Ontario. The latter were likely exchanging maize for skins and hides with their northern Algonquian neighbours as the Wendat were known to do in the historic period; the first Europeans in the area described the Wendat economy and their growing of a surplus of corn for trade as representing the granary of the north.

No agriculturalists actually arrived in the Georgian Bay region, however, until the late thirteenth century. These were some of the Iroquoian-speaking ancestral Wendat from the north shore of the

16th century *maize cobs, Mantle site.*

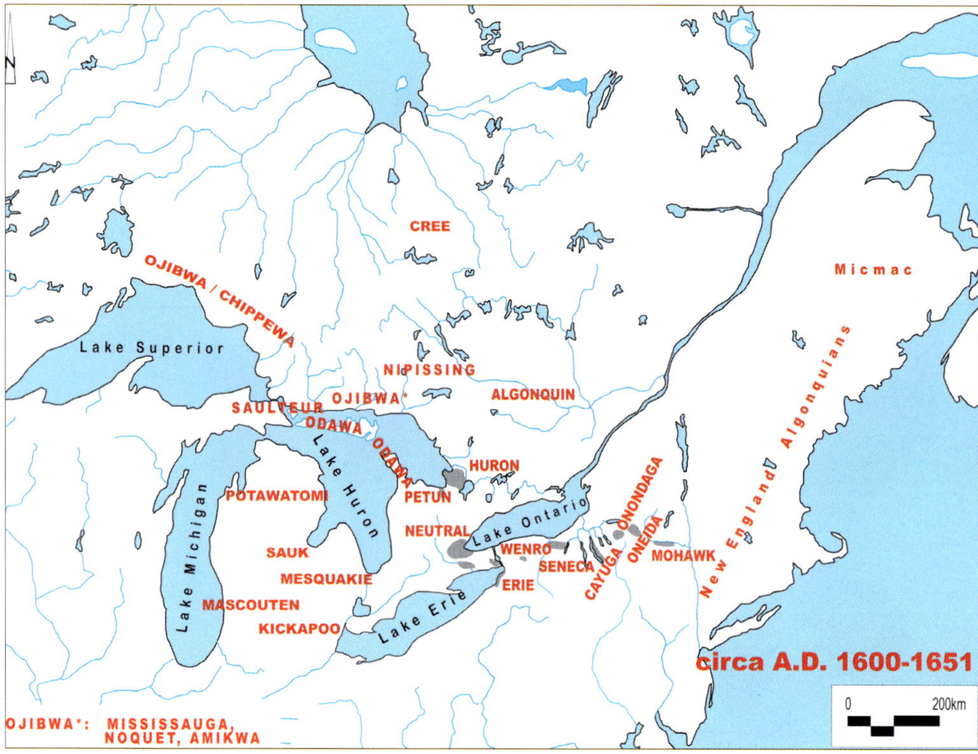

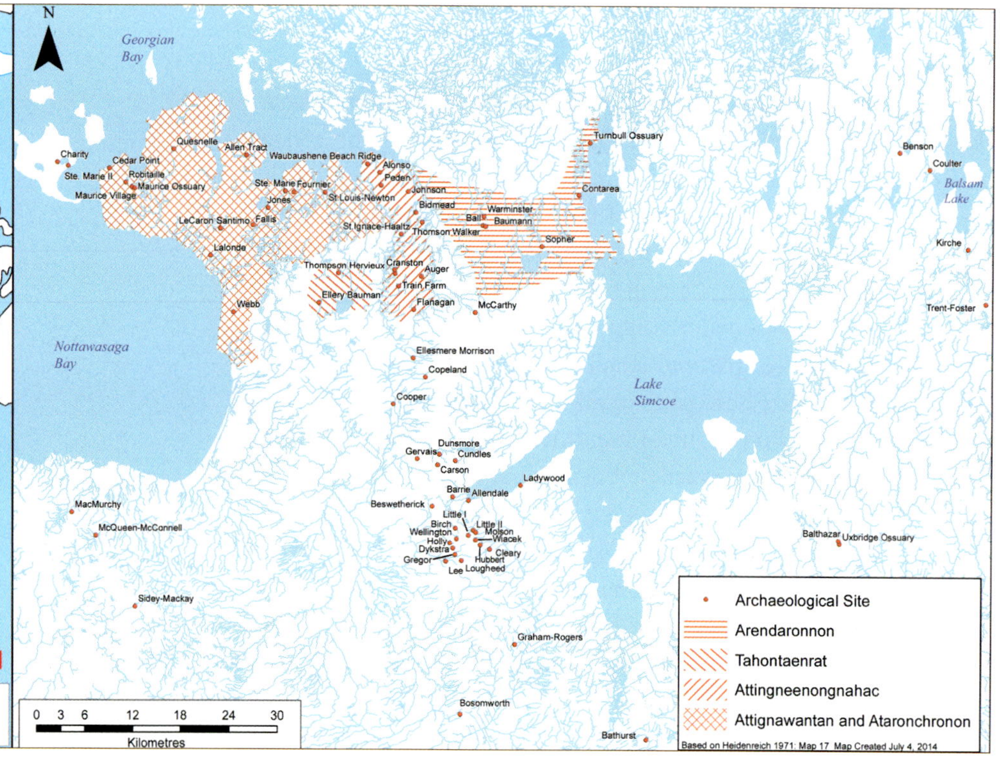

Left: **Early 17th century** location of Indigenous nations .

Right: **Early 17th century** Wendat tribal locations.

Lake Ontario region who had likely negotiated their arrival with northern Algonquian bands who resided in the area, especially those who used the Simcoe County landscape. Some thirteenth and early fourteenth century sites in the Kempenfelt Bay area yielded clear evidence of co-habitation by Iroquoian and Algonquian peoples. The best evidence for such a scenario includes the presence of American eel at the Wellington site, likely representing food brought by the ancestral Wenda, from the north shore of Lake Ontario since it was unavailable locally; significant differences in the frequencies of chert types between two house structures on the site, one yielding significant quantities of chert thought to be far more available to local Algonquians than ancestral Wendat and the presence of ritual burials of multiple small fur-bearing species, also documented in four features at the nearby, slightly later Holly site. These features are unique finds in Southern Ontario, and are similar to the interment of disarticulated, generally young or immature dogs (and other animals) in ceremonial contexts among Algonquian-speakers of the region.

By 800 years ago, people residing east of the Bruce Peninsula and throughout Simcoe County were fully committed agriculturalists eating more than a pound of maize per day, comprising more than half of their diet. Maize was supplemented by beans and squash; sunflower and tobacco were also grown by then.

An absence of deer from Simcoe Couny at this time meant that hunting parties were sent south at the same time as hides were sought through exchange with their Algonquian neighbours. Wendake, the home of the Wendat, was bounded by Lake Simcoe to the east and Matchedash and Nottawasaga Bays at the southeastern corner of Georgian Bay to the west. Their confederacy consisted of four allied nations: the *Attignawantan* (Bear), *Attigneenongnahac* (Cord), *Arendarhonon* (Rock), and *Tahontaenrat* (Deer). Another population known as the *Ataronchronon* (Bog) does not appear to have been an independent member of the confederacy and were a division of the *Attignawantan*. Their collective name for themselves, *Wendat*, has been interpreted as meaning "islanders" or "dwellers on a peninsula." More specifically, the Bear occupied the southern shore of Georgian Bay, along the Penetang Peninsula from Nottawasaga Bay to Midland

Bay while the Ataronchronon occupied the south shore of Georgian Bay from Midland Bay east to the Coldwater River.

After the arrival of Europeans, there is a rich seventeenth-century documentary record of the lives of the Huron-Wendat. The three principal sources are the works of Samuel de Champlain, an experienced soldier and explorer who recorded his observations of Wendat (and Tionontaté) life, during a winter spent among them in 1615–16; the detailed account of Gabriel Sagard, a Recollet friar who spent the winter of 1623–24 with them; and the annual accounts of the Jesuit priests who lived among the Wendat from 1634 until 1650.

The Tionontaté lived immediately southwest of the Wendat in the Collingwood-Craigleith area having migrated there a few decades prior to European contact. Their economy was similar to that of their Wendat neighbours; they also enjoyed privileged access to the items secured by the Odawa through their extensive trade networks. Their confederacy included two separate groups, the Wolf and Deer. At the time the Jesuits arrived in Huronia, the Wendat and Tionontaté were allied against common Iroquois enemies. Their combined population prior to the spread of European epidemics in the 1630s has been estimated to have been 30,000. The ancestral Tionontaté and Wendat Iroquoian-speaking horticulturalists were closely tied economically and politically to their Anishinaabeg neighbours. This becomes most evident during the contact period from French documents that discuss the close connections that the Tionontaté had with the Odawa and the Wendat with the Nipissing.

In 1640, the Jesuits who were living with the Wendat recorded that along the eastern shore of Georgian Bay, there were three Anishinaabeg groups called the Ouasouarini, Outchougai, and the Atchiligouan or the Birch Bark people, the Heron people, and the Black Squirrel. Along the north shore of Georgian Bay was the Amikouai (Ojibwa), or nation of the Beaver, the Oumisagai (Mississauga), and the Baouichtigouian or people of the Sault. Throughout the 1640s, the Sault people are documented at the juncture of Lake Huron and Lake Superior, near what is now Sault Ste. Marie.

The Nipissing who inhabited the region of the lake by the same name were situated on the historic route between Quebec and the Wendat country. In the autumn, some Nipissing made the journey south to spend the winter with the Wendat.

The Odawa or Ottawa were situated on the Bruce Peninsula, Manitoulin Island, and the eastern shore of Georgian Bay. When Samuel de Champlain arrived at the mouth of the French River earlier in 1615, he became the first European known to see Georgian Bay and encountered a large contingent of Odawa warriors who he called Cheveux Relevé (High Hairs), so named for their distinctive manner of wearing their hair raised in the front in a style and described by the French as a pompadour.

There are a number of examples of shared ceremonies between the Iroquoian and Algonquian nations. A Feast of the Dead, for example, was held in September, 1640 about 20 leagues or 60 kilometres north of Wendake along the Georgian Bay shoreline. Such ceremonies had been common among the Huron-Wendat for several

Early 18th century *drawing of a Wendat Feast of the Dead, Joseph François Lafitau.*

Left: ***Semi-subterranean sweat lodge**, Wiacek site.*

Right: ***Wiacek site** sweat lodge with bear skull.*

centuries but were new to the Anishinaabeg. At the time of village relocation, those that had died over the tenure of a village would be gathered to be interred together in an ossuary thereby creating a community of the dead. In this case, it is estimated about 2,000 people from many Nations assembled for the gathering. The Jesuit observers singled out the Nipissings, who appear to have been the hosts, and noted the presence of the people of the Sault, and also the Wendat. It was apparently a multi-day ceremony with feasting, mourning, games, and gift exchange. The Jesuits also described a splendid Feast that was prepared for the Wendat and noted that the Algonquin Nations were served separately from the Wendat as they spoke different languages. Separate ceremonies were subsequently held for the preparation of the dead and associated gifts.

Community Building

Throughout the fourteenth and fifteenth centuries, ancestral Wendat people in Simcoe County inhabited large, generally unpalisaded villages of four to six longhouses, situated adjacent to a source of fresh water and surrounded by hundreds of hectares of well-drained loam soils on which to grow crops. Given the ever-declining fertility of the corn fields and supplies of firewood, as well as the problem of managing considerable quantities of organic and inorganic refuse, these villages were only occupied for a few decades. The Wendat field systems were so extensive that the Jesuits, a century later, speak of getting lost in the fields while travelling from village to village. As the fertility of the fields were depleted, it was necessary to clear new land until after 15 to 20 years the distance to travel to outlying fields from the village became so great that it was necessary to establish new villages.

There were dozens of fourteenth- and fifteenth-century Wendat villages throughout Wendake where chiefs were elected from the principal clans. Villages housing hundreds if not more than a thousand people required new social and political structures to regulate village affairs and relations between villages. Village councils, more formalized community planning, various social groups such as curing societies, as well as group rituals such as feasting and community burial, all emerged during this time period. Curing societies are first visible in many of the longhouses in the form of semi-subterranean sweat lodges – a special type of feature that made its appearance in the late thirteenth century. These structures were likely used for religious and medicinal purposes, as the practice of sweating is thought to have been a means by which to communicate with the spirit world. They consisted of large, deep, rectangular pits with entrance ramps at one end, extending upwards to ground level. Archaeologists can identify the bottom of the pit by a thin, flat layer of dark soil mottled with charcoal and ash containing fire-cracked rock and artifacts. This layer, the living floor, developed by the repeated dousing of hot rocks to produce steam. They were covered by a frame made of bent saplings, perhaps covered with hides or bark. Sweat lodges may have served to cement the newly emerging social and political ties among the various members of the village. Occasionally, an animal skull is found on the living floor of a lodge, perhaps symbolizing clan identity. Elaborate bone tools and pipes have also been found in these features, which sometimes served as places to bury the dead. The partially excavated Hubbert site, west of Kempenfelt Bay, contained at least three houses featuring a total of seventeen semi-subterranean sweat lodges indicating considerable effort at social and political integration at the site.

In theory, holes in the roofs of longhouses allowed smoke from the hearth fires to escape although Wendat health suffered as a result of spending too much time in smoke-filled longhouses. Not only were respiratory infections common but some also contracted tuberculosis. Living in close proximity to organic refuse along with rodents, dogs, and considerable numbers of people would have heightened susceptibility to disease.

By the mid- to late fifteenth century, some villages attained sizes of two or more hectares resulting from population growth and the coalescence of smaller villages into larger ones. Attignawantan (Bear) and Attigneenongnahac (Cord) were the original co-founders of the Wendat confederacy since both had been resident in Wendake for at least 200 years previously. Later additions to the confederacy were Arendahronon (Rock) who moved into Wendake about 1590 and the Tahontaenrat (Deer), who joined around 1610. Champlain was told by the Arendahronon that they had formerly lived in the Trent Valley and had abandoned the area due to fear of enemies. The Tahontaenrat (Deer) likely originated with north shore of Lake Ontario communities. The Tahontaenrat and Attignawantan spoke different Wendat dialects perhaps attesting to their geographic separation. Thus, by shortly after the turn of the seventeenth century, the north shore of Lake Ontario was devoid of permanent settlement, these populations having relocated north to join the Wendat and/or Tionontaté confederacies.

For the next four decades, French traders and missionaries travelled from Quebec to Wendake usually taking the same route via the St. Lawrence to the Ottawa River to the Mattawa River to Lake Nipissing to the French River to Georgian Bay. The route taken from the mouth of the French River to Wendake followed the east side of Georgian Bay to avoid the Iroquois or Haudenosaunee, enemies of the Wendat and their allies.

The Onset of Disease

By the 1620s, there were several well-fortified Wendat villages which housed hundreds of warriors who were prepared for Iroquois raiders; most nations also had ancillary villages as well as nearby locales where Algonquians came to winter. According to seventeenth-century accounts, the Wendat-Tionontaté population totaled 30,000 to 35,000 before the initial epidemic of 1634. The 1630s and 1640s were disastrous times for the Wendat. In 1634, measles spread throughout the Attignawantan villages during the winter followed by influenza in early September 1636, persisting until spring of 1637. Between 1634 and 1637, Wendat and Tionontaté populations experienced a

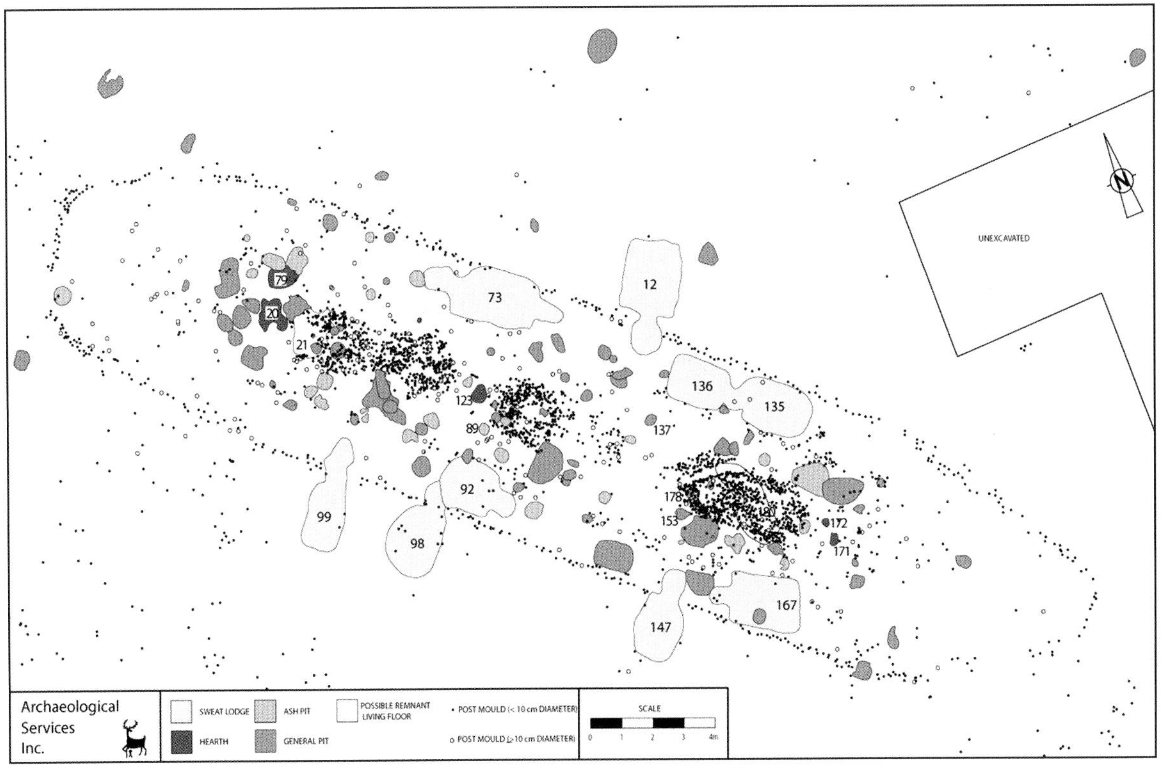

Longhouse *with sweat lodges, Hubbert site.*

Water colour of the ruins of Sainte Marie II by Father Felix Martin.

20% decline, leaving just 23,000 people alive by the end of 1637. An epidemic of smallpox ravaged the Wendat and Tionontaté between early fall 1639 and spring 1640, reducing their population to between 10,000 and 12,000 as documented by the Jesuits in the 1639-1640 census. In 1639, the French established a central mission that became the first European settlement on the Great Lakes. Named Sainte Marie by the Jesuits, it was located in Attignawantan (Bear) territory just south of Georgian Bay on the Wye River in present-day Midland. The settlement became the primary centre of Jesuit missionary activity in Wendake. Sainte Marie has been recreated to provide visitors with a glimpse of what life may have been like during the first half of the seventeenth century.

Over the next ten years, the Wendat were attacked repeatedly leaving only 15 settlements remaining in 1649. Some Wendat fled to the Tionontaté, but in December 1649, their main fortified village of Etharita was destroyed by Iroquois and about 1,000 people were forced to travel to Iroquois country. Another 500 to 1,000 Wendat-Tionontaté fled Tionontaté country to settle on Gahoendoe (Christian Island). With the escalation of hostilities with the Iroquois, the Tahontaerat left in 1649 to reside with the Neutral and then eventually with the Seneca. In 1648 and 1649, three villages near to the mission of Sainte Marie fell to the Iroquois including Taenhatentaron (St. Ignace), the site where Jean de Brébeuf and Gabriel Lalemant were martyred.

The most comprehensive excavation of a site of this period (and its reconstruction) occurred at Sainte Marie. By 1648-49, the presence of the Jesuits and their lay assistants in Wendake had increased substantially to around 50 Frenchmen, coincident with the growth of the Sainte Marie mission into a well-fortified French settlement and associated farm. Detailed information about the layout and buildings within the mission were determined through a series of archaeological investigations by Ken Kidd of the Royal Ontario Museum and subsequently by Wilfrid Jury of the Museum of Indian Archaeology at the University of Western Ontario. These revealed a wooden palisade and internal ditch complex as well as later stone fortifications along with subdivided European and Indigenous compounds, the former with a complete longhouse and the latter with a chapel, trade shops, a cookhouse, barracks, a barn for domesticated animals, a hospital, and various other dwellings.

Wendat refugees from Iroquois aggression from the rest of Wendake, including the widows and orphans from Ossossané, fled to Gahoendoe (Christian Island) in the spring of 1649. When the advance of the Iroquois had reached the gates of Sainte Marie I, the remaining Wendat and Frenchmen departed with all their belongings by boat and raft, eventually reaching the south-eastern shore of Christian Island. While an exact number is unknown, thousands had fled to the island. The Jesuits and their workmen constructed a four-cornered fort with stone curtain walls and bastions named Sainte Marie II and also helped to strengthen the fortifications of the adjacent Wendat village. Conditions at the settlement were disastrous due to disease and crop failure brought on by drought and famine, which in turn led to cannibalism. There was also the constant threat and harassment by the Iroquois. The French and about 300 remaining Wendat abandoned the fort for Quebec in June, 1650, paddling up the shores of Georgian Bay to the French River and their route to Quebec, ending the French and Wendat occupation of historic Wendake. The first recorded visits to Sainte Marie II subsequent to the French period appear to have been in the nineteenth century when various clerics recorded their observations of the ruins of the fort.

The Tionontaté had fared no better at the hands of the Iroquois and abandoned their territory in 1649. Several Tionontaté sites have been investigated including the Plater-Martin site, located on a terrace overlooking Georgian Bay in Craigleith. Well known for well over a century, the site was investigated by famed local archaeologist Charles Garrad, who identified the site as the Tionontaté village of Ekarenniondi, which was also known as the Jesuit mission of St. Mathew. Plater-Martin and the nearby Plater-Fleming site are believed to have been contemporaneous, comprising the Jesuit Mission of St. Matthias. Both villages date to the last decade of Tionontaté occupation prior to their dispersal by Iroquois in 1649. Such a late date for these sites is supported by the recovery of relatively large numbers of European trade goods, including European glass beads, iron tools, brass kettle fragments, gunflints, and Jesuit rings.

According to the Jesuits, the village of Ekarenniondi, while principally Tionontaté, was also occupied by Odawa. The presence of bear jaw tools at the site, typically found on Odawa sites throughout the Georgian Bay region, supports Champlain's observation of Algonquian inhabitants at the site and reflects the close relationship between the Tionontaté and Odawa, which had been first observed by Champlain three decades earlier. The site also yielded evidence of a unique smoking pipe industry, represented by finished pipes some with animal and human figures carved on the bowls and numerous pipe preforms all made from local limestone. Large quantities of marine shell, which originated on the mid-Atlantic coast, were found on the site. Notable is a piece of conch columella which has been shaped into a gorget and was likely worn around the neck by threading a thong through the two drilled holes. Pieces of the shell would have been obtained through a trading network that extended to the southern Atlantic coast. Tubular marine mollusk beads and a well-made tubular "wampum" bead, perhaps manufactured from quahog shell, were also recovered.

The Plater-Fleming site situated within 400 metres of Plater-Martin is estimated to be just over one hectare in size with several longhouses surrounded by a three-row palisade. Five dog burials were encountered, each showing signs of ritual dismemberment, consistent

Plater-Martin and Fleming goods

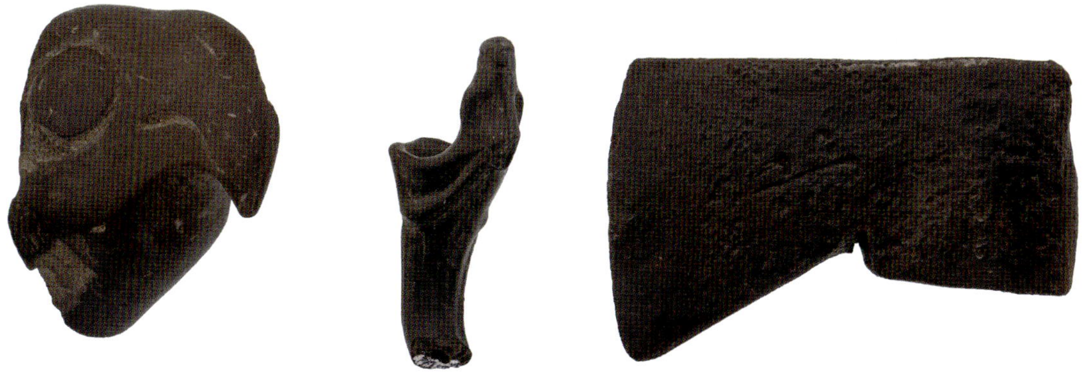

Plater-Martin and Fleming goods

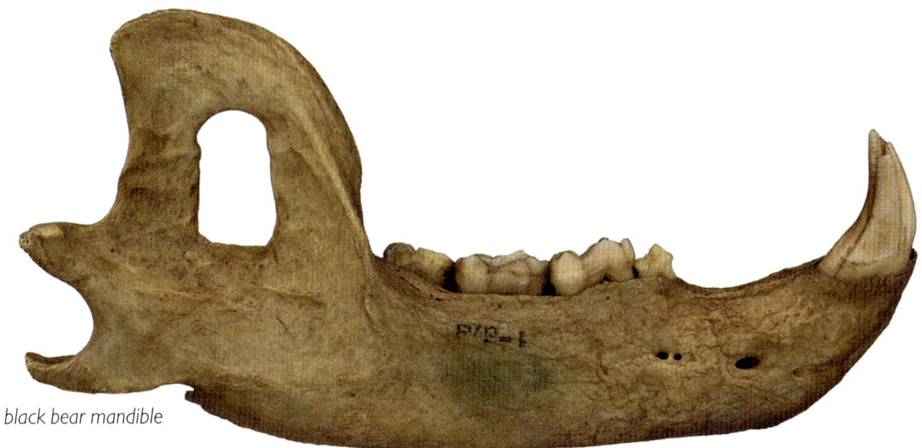

Plater-Martin black bear mandible

Courtesy of John Howarth, courtesy of ASI

with Anishinaabeg ceremony. Such ceremonies involving the ritual treatment of dogs are known from the documentary record of the Upper Great Lakes, as well as being identified archaeologically on sites in the Bruce Peninsula, Manitoulin Island, and on the Canadian Shield. Unique to the site is a bear effigy pipe made of steatite. Among the recovered animal bones from the site, as well as from Plater-Martin, were modified black bear mandibles associated with Odawa bear ritual.

Several other archaeological sites related to the Odawa have been investigated in and around Georgian Bay such as the Providence Bay and Shawana sites on Manitoulin Island, Hunter's Point site on the Bruce Peninsula, and the Glen site on Flowerpot Island. Common to all these sites is a wide variety of cherts used for making stone tools such as projectile points, knives, and scrapers. These varieties include Fossil Hill, Wike, Detour, Norwood, Kettle Point, Hudson Bay Lowland, Onondaga, and Bayport, most of which can be found within the Georgian Bay-Lake Huron watershed but would have required long-distance travel by canoe to obtain. At the Providence Bay site, bone and shell artifacts including beads, sucking tubes, and shell pendants were found as well as European items such as iron knives, trade axes, and trade beads. Maize was grown and beaver was the most prominent mammal followed by snowshoe hare, fox, domestic dog, raccoon, marten, fisher, mink, black bear, white-tailed deer, woodland caribou, and moose. Fish formed the major animal component of the diet and included sucker, trout/whitefish, lake sturgeon, pike/muskellunge, catfish/bullhead, burbot, bass, yellow perch, and yellow walleye. Also recovered were several species of bird, with loon being the most prominent, as well as several varieties of turtles. Of special interest was the presence of burials of dogs, beaver, loon, and eagle.

The Hunter's Point site is a large multicomponent Odawa site located on the west side of Georgian Bay on the Bruce Peninsula. It is known to have special spiritual and cosmological significance for Indigenous peoples in addition to its use as a hunting and fishing camp. In fact, the abundance of fish and birds in the region may be the reason why the area developed as a place for healing. The presence of bear, dog, hawk, and eagle is significant given their affiliation with ceremonies. While the main occupation has activity areas dating to between 1600 and 1625, based on the presence of European trade items, the site has also yielded evidence of ceremonies performed by late-nineteenth-century Mississauga, Ojibway, or Potawatomi indicating continuity in the traditional use of the site for hundreds of years from the pre-contact past to the present. Ceremonial objects found at the site include a pendant made from dark-reddish brown shale depicting Anishinaabeg iconography relating to cosmological beliefs. One side of the pendant is depicted with Mizipichu, another land-based animal and radiating lines representing Megis. The other side of the pendant depicts a shaman and power lines. A shell amulet resembling a Thunderbird was also found.

The Glen site, located on Flower Pot Island, is an Odawa fishing camp occupied during the early seventeenth century and was most likely inhabited between October and November during the spawning season of various fish. Some of the Lake trout remains were exceedingly large by today's standards-with up to 100 pounds.

Conflict and Change after 1650

By the beginning of the 1650s, the southern Georgian Bay region had been temporarily abandoned by the Wendat, Tionontaté, and their Algonquian allies due to decimation by European-introduced disease and attacks by the Iroquois. In the 1660s, the Iroquois established a series of settlements in Southern Ontario at strategic locations along the canoe-and-portage routes that linked Lake Ontario to Georgian Bay and the upper Great Lakes, through Lake Simcoe. By the 1690s, these settlements were abandoned by the Iroquois and the Mississauga had moved into the territory and was the Nation with which the British eventually reached land agreements with in the mid-eighteenth century, although many of these remain contested.

Within a decade, however, the former inhabitants of the Georgian Bay region were travelling across their territory and even establishing settlements. Pierre Esprit Radisson, for example, recorded in his travel journals the presence of an Odawa village also inhabited by some Tionontaté-Wendat, on Manitoulin Island around 1658-1660. Their group also engaged in hostilities with an Iroquois party, and Anishinaabeg oral tradition holds that they had a major role in finally pushing the Iroquois out of Southern Ontario in the late 1600s. They speak of a huge pincer movement with warriors travelling down the east side of Georgian Bay, from Sault Ste. Marie, while others came across from the Bruce Peninsula from Manitoulin Island, and still others from farther south. Numerous battles are related including ones near the current Saugeen Reserve on the shore of Lake

Huron, in Wasaga Beach near Collingwood, on Indian Hill near the Teeswater River at Owen Sound, on White Cloud Island in Colpoys Bay, on Griffith Island at Cabot's Head, in the Fishing Islands around the Bruce Peninsula, and at Red Bay, the latter having purportedly received its name from the bloody waters following the defeat of the Iroquois at that location.

There are also historic accounts of Anishinaabeg within the Georgian Bay region in the early eighteenth century; for example, the "Enumeration" of 1736 places people on Manitoulin Island, all from the Crane branch and others at Matchedash Bay, Lake Nipissing, and at the Falls of Saint Marie. This same Matchedash band was recorded later living on Georgian Bay Islands from 1838 to 1842. Still others resided along the Spanish River, on the north shore of Georgian Bay. In the early 1760s, Alexander Henry encountered a village on the island of La Cloche on the north side of Manitoulin Island and in 1793, John Simcoe met Indigenous peoples at the mouth of the Severn River. Indigenous settlements on the Bruce Peninsula also appear on maps of the period, first on a 1725 map of "Canada" by Chaussegros de Léry and on a 1788 "Sketch of Lake Huron" by Gother Mann. Cape Croker Indian Reserve ("Neyaashiinigmiing" or Nawash) means "point of land surrounded on three sides by water." They trace their history through centuries of European encroachment and a series of surrenders to the reserve they inhabit today.

The Present Day

By the turn of the nineteenth century, there are also clear historical records as well as oral accounts of Anishinnabeg families hunting, fishing, farming, and trapping throughout western Muskoka on the eastern shores of Georgian Bay. Chief John Aisance, for example, led his band to Beausoleil Island in 1842 from the Coldwater settlement. They established a small largely agricultural community with settlements at Cedar Springs and further north in the Beausoleil Bay area. The Cedar Springs settlement consisted of 20 log houses, a barn, and a schoolhouse. Crops were grown on the surrounding islands and more than 5,000 pounds of maple sugar were produced annually. Due primarily to poor soil fertility, the band moved to Christian Island in 1857. The Cedar Springs settlement is now an archaeological site consisting of 25 artifact concentrations, occurring in association with house mounds or discrete scatters; the second settlement contains in addition to the nineteenth-century occupation, pre-contact components dating to 3,000 years ago.

Chief William Yellowhead's Anishinaabe name, "Mesquakie," is the basis for the word Muskoka, which was used by early fur traders to refer to the Muskoka River watershed. Ojibwa families living on the Indian (Biasong) River were later referred to as the Muskoka band, not because of any connection with Mesquakie, but because of their geographical location. They lived in the area of Port Carling until 1868 when, due to the encroachment of Euro-Canadian settlement, they were compelled to move to Parry Island (now Wasauksing First Nation). The Menominee family was also well known to early settlers and surveyors during the middle- to late-nineteenth century and they too had moved to Parry Island in the early 1870s. It was not until 1917 that the Moose Deer Point Reserve was established; that band had also been at Beausoleil Island from the mid-nineteenth century onwards.

The southern portion of Muskoka is still familiar territory to M'ngikaning (Rama) families. The Severn River was an important travel route to hunting territories in the interior of Muskoka and along Georgian Bay. As late as the 1930s, the Simcoe, Williams, and Ingersoll families were living, at least for part of the year, on Sparrow Lake.

More generally, traditional travel routes would have involved overland trails and canoe routes marked by blazes and natural orientation features, but included portages and transitory camps. In Muskoka many of the major canoe routes and portages of the past continue to be used today by recreational canoeists. The canoe route from Georgian Bay to the Huntsville and Lake of Bays area was an important route for Muskoka Band members. The Severn River also represented an important east-west canoe route providing access to Georgian Bay and points east via the Black River.

The late nineteenth century saw the settlement of other Indigenous groups on the eastern shore of Georgian Bay. A large tract of land near Georgian Bay was purchased and settled in the 1880s by Mohawks originating from Lake of Two Mountains near Montreal. The Wahta Mohawks were no strangers to this territory, however, having travelled extensively through the region since the eighteenth century. Not only were they familiar with the area through their participation in the great fur trade canoe brigades, but they also ran fur trade posts and acted as guides for Euro-Canadian travellers and explorers. Established first as the Gibson Reserve in 1881, it is now known as the Wahta Mohawk First Nation. The reserve, situated near Bala, sits astride Highway 400 and is known for its successful cranberry harvesting and maple syrup enterprises.

Manitoulin and the surrounding islands are also central to the story, both past and present, of Georgian Bay. Manitoulin means "Spirit Island" in the Ojibway language and is regarded as sacred to Anishinaabeg peoples. As we have seen, the Island has been inhabited for at least 10,000 years and was intermittently occupied during and after the Iroquois wars. In the 1830s, Lieutenant Governor Bond Head requested the agreement of those living at the time on Manitoulin Island to permit other Nations to move to there to form a protected homeland reserved exclusively for those First Nations. A treaty to that effect was signed in 1836 (The Bond Head Treaty), which was subsequently ratified by the Crown. All of the islands in Georgian Bay, and as far west as Sault Ste. Marie, were reserved for First Nations—the Anishinaabe were guaranteed the "innumerable fishing islands" and the "Great Father [would] withdraw his claim to these islands." In 1862, however, the government was concerned that few other nations had moved to Manitoulin, which resulted in renegotiation of the treaty and the opening up of the island for settlement by non-Native people. The Wikwemikong chief did not accept this treaty, and that reserve remains unceded (land and sovereignty was not surrendered to the Federal Government). The island to the west of the Wikwemikong territory was ceded to the Crown, and Native Reserves were established for the various Aboriginal settlements.

At present day, there are seven reserves on Manitoulin including Wikwemikong, M'Chigeeng (formally West Bay), Sheguiandah, Sheshegwaning, Wauwauskinga, Zhiibaahaasing (formerly Anishinabek of Cockburn Island), and Aundeck Omni Kaning (formerly Ojibways of Sucker Creek) First Nations, all of which not only provide a 10,000- year-old link to the history of Manitoulin Island, but all of Georgian Bay. The presence of these and the other Georgian Bay reserves attest to not only the founding population of the region but to a series of vital communities guarding both the integrity of their past and their future on the sacred landscape that they too call home.

Highway 6 meanders *through the white quartzite foothills of the La Cloche Ranges, a 2 billion year old fold and thrust belt.*

Glacially-streamlined rock of the Canadian Shield near Whitefish Falls.

CHAPTER

9

WHERE ROCKS RULE: THE SETTLER EXPERIENCE IN GEORGIAN BAY

NICK EYLES

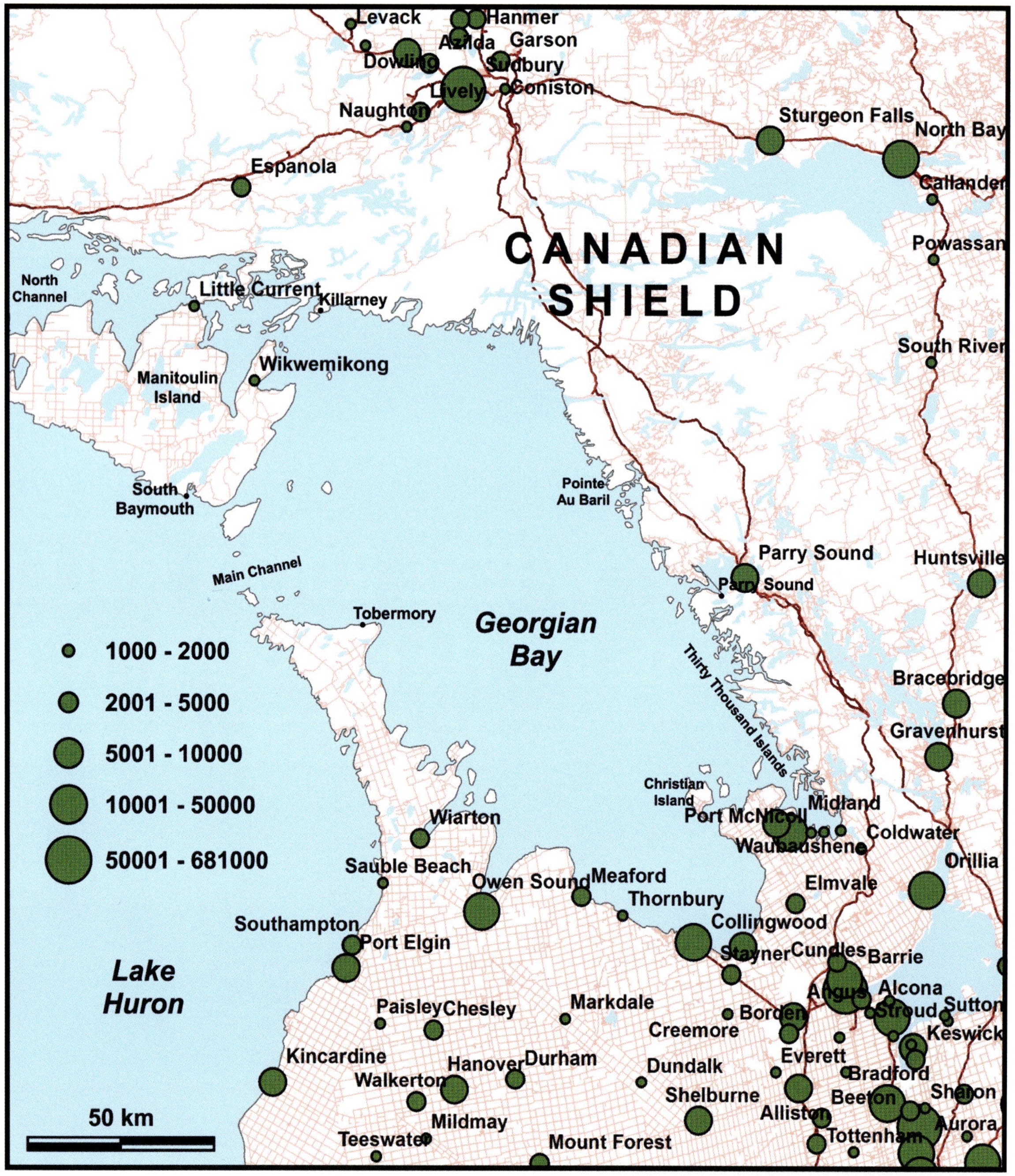

***The densely settled south vs. the sparsely settled north** reflects the geology of Georgian Bay and the lack of soils on the Canadian Shield.*

Where rocks rule: the settler experience in Georgian Bay

Georgian Bay has always been a geological frontier a virtual 'land in between' along the border of very different geological realms. It is not surprising that the recent human story of Georgian Bay after the arrival of European settlers has been dictated by rock.

The hard rocks of the Shield to the north and east, contrast with the much younger soft sedimentary rocks of the south and west. Migratory hunting, gathering and fishing across this boundary was a fundamental part of Aboriginal subsistence and had prevailed for thousands of years on the forested lake-studded Shield, with more recent sedentary farming of rich soils to the south both by First Nations and later European settlers; as dramatic a contrast in landscape and resulting human activity as can be found anywhere in Canada. The interchange of goods across the border was the foundation of the fur trade which defined early Canada in the eyes of the world after Samuel de Champlain first witnessed the bartering of corn for northern furs somewhere near Lake Simcoe in 1615. It is not surprising that the first European settlement was built on the southernmost margin of Georgian Bay (Sainte-Marie, near present day Midland) in 1639 as a mission-fort facing lands and waters 'unknown to the north.' Georgian Bay was a military frontier during the War of 1812; contested by Great Britain and the United States and not really wanted by either for whom it was of little value other than as an empty buffer.

The sparse population and sporadic road network of the lake-studded Canadian Shield east of Georgian Bay, along with the limestone plains of the Bruce Peninsula and Manitoulin Island on the west, contrasts greatly with the more densely settled croplands and systematic grid of concession roads to the south. The famous Ontario geologist A.P. Coleman wrote in 1913 of the fundamental difference between the two regions: 'After passing the boundary .... the traveller is struck by the change of scenery, rounded hills of reddish gneiss rising irregularly above valleys occupied by a lake; for the region stretching two thousand miles to the north of the Paleozoic border is typical rocky lake country, with thousands of rock-rimmed bodies of water.'

The year 2015 saw the 400th anniversary of the exploration of eastern Ontario, including the Georgian Bay area, by Samuel de Champlain (1567-1635). He was many things, soldier, colonial administrator and statesman primarily, but at heart he was essentially a map maker to the French King Henry IV who was fascinated by the emerging geography of the 'New France.' Champlain was known as 'géographe du roi' or 'Geographer to the King' having explored the eastern side of Georgian Bay, what is now known as the Thirty Thousand Islands, in 1615 and made a map of 'Huronia' in 1616. This showed he had been unaware that he was on the eastern embayment of a much larger lake now known as Lake Huron. Champlain travelled with a group of Huron-Algonquin warriors who in exchange for help against their Iroquois enemies assisted Champlain in exploring their territory west of the Ottawa River. They travelled by canoe from Lake Nipissing west along the French River to Georgian Bay commenting on the great size of the fish in the Bay and the 'great quantities of animals and birds.' He was however unimpressed by the terrain surrounding the east coast of Georgian Bay being 'rocky and hilly' and lacking in any arable land. His party was forced to subsist on blueberries and strawberries to avoid being reduced 'to straits.' It is scarcely surprising then that he was struck when leaving southern Georgian Bay via Matchedash Bay and Penetanguishene by the 'great change in the country' and the large population (he estimated 30,000) living in large palisaded villages surrounded by corn fields; which he described as 'fine and fertile… and very pleasant.' Later that year Champlain and his party moved south through the Kawartha Lakes (down what is now essentially the Trent Severn Waterway) to Lake Ontario and received two arrow wounds in combat with the Iroquois at a fortified village a few miles south of Oneida Lake in what is now Upper New York State. He was forced to overwinter in Huronia, just north of Lake Simcoe in 1615-16 where he made many observations of local life and customs. He left Ontario in May 1616, never to return.

Georgian Bay largely escaped the great conflict of the war of 1812. The closest the war came was in mid-August 1814 when the British-captained (Lieutenant Robert Worsley) HM Schooner Nancy a former fur trading vessel, became trapped in the mouth of the Nottawasaga River in southern Georgian Bay by US forces who were seeking to cut British supply lines to Mackinac Island which lies at a crucial strategic position where Lake Michigan meets Lake Huron and which the Americans had earlier tried to take unsuccessfully, deciding to blockade it instead. The island was the key to controlling northern Lake Huron and the English-dominated

***Memorial plaque** to Champlain on Frying Pan Island, near Sans Souci, Georgian Bay.*

fur trade. The Nancy's crew burnt the ship lest if fall into enemy hands escaped and rowed back to Mackinac Island where they exploited an opportunity near Drummond Island to capture two of the US schooners (Tigress and Scorpion) that had earlier discovered the Nancy in the Nottawasaga River. Pressed into British service by the Royal Navy the two vessels were later based at the Naval Establishment on Lake Huron at Penetanguishene now known and operated as Discovery Harbour. A replica of a schooner of the time HM Tecumseh was built and is on display there, while the wreck of the original Nancy preserved in silt and discovered in the 1920's can be seen at the Nancy Island Historic Site near Wasaga.

With the arrival of the railways along the fertile southern coast of Georgian Bay in the 1850's small lakeshore communities of Owen Sound, Collingwood and Midland were transformed into busy Great Lakes transhipment ports for grain, fish, people and timber. To the east, the railroads skirted inland of its eastern rocky island-strewn

Mural in Parry Sound *"The Brag Load" by Brian Romagast.*

coast and apart from small railway stations such as Key Harbour, Pointe au Baril and the community of Parry Sound, the railway boom had little real effect in bringing that part of Georgian Bay into the orbit of the urban south. Agriculture was difficult to impossible and it was a place to travel through and not settle. The area's lack of prospects were underlined by the collapse of the short-lived lumber industry with the cutting of the last forest stands in the late nineteenth century followed by that of the fishing industry after more than a century of heavy overfishing.

Georgian Bay gets named

First charted in 1815 by Captain William Owen, Georgian Bay was named in 1822 by Lieutenant Henry Wolsey Bayfield of the Royal Navy after King George IV. Earlier, Samuel de Champlain originally named it 'La Mer Douce' (the sweet water sea) when exploring it in 1615-16; an early map also portrays it as Karegnondi derived from 'lake' in the language of Petun First Nation who lived along its southern shores.

Ownership of the coastlands and waters later to be called Georgian Bay had been settled by the War of 1812 between Great Britain and the USA when the Treaty of Ghent (1814) kept it in Canadian hands. The first detailed cartographic representation of the Bay was made in 1828 when the British Admiralty published Lt. Henry Bayfield's maps that noted its dominant rocks of gneiss and granite. Bayfield's work was based on an arduous four year survey from 1819-1822 among the 'seas of Lakes Huron and Superior … as heavy as those of the Ocean', a land where clouds of mosquitos 'literally darkened the air' and many a night was spent in near zero temperatures 'after many very narrow escapes when caught in heavy gales late in the Season… with the ground covered with a foot or two of snow.' Bayfield considered that 'there has perhaps been no survey made of greater labour of detail, than that of the 20,000 Islands, Islets and Rocks of Lake Huron.' The northeast shore of Georgian Bay was singled out as an area to avoid; thousands upon thousands of seemingly chaotic red and grey granite knobs rising from the waters of Lake Huron to form an almost impenetrable barrier of treacherous rocky shoals and reefs. The Canadian Shield though named for its relatively high standing topography in the middle of the country, was also aptly named; it was indeed a barrier to penetration.

Embraced but not really loved

The Robinson Treaties of 1850 formally brought Lake Huron and Georgian Bay into the sphere of Canada West (what is today southern and eastern Ontario) creating the vast Ottawa-Huron Tract across a large area of the Canadian Shield. It was opened up for business and settlement by Europeans by moving and relocating First Nations onto fixed reserves where they surrendered any interest in the eastern and northern shores of Lake Huron. These reserves were placed in areas that were considered to be wilderness by settlers supposedly befitting the lifestyle of its first inhabitants. By the stroke of a pen, aboriginals gave up their long-lived spiritual and cultural links with the land.

But what exactly had they given up? A 'barren waste,' 'sterile,' 'unfit for settlement', 'a broken land' and 'difficult and dangerous access' are repeated in the reports and correspondence of the time. Commander J. Boulton who commenced a hydrographic survey of Georgian Bay in 1894 spoke of the 'notorious northeast shore of Georgian Bay' choked with numerous uncharted shoals where the mouths of the main rivers disappear into a myriad of granite islands. He wrote 'this shore possesses all the characteristics unfavourable to

Above: ***Much of eastern Georgian Bay*** *was described as unfit for settlement in the nineteenth century.*

Left: ***Once seen as a sterile rock-strewn wilderness,*** *eastern Georgian Bay is a unique aquatic and terrestrial ecosystem.*

the hydrographical surveyor. In the first place, the coast has by some mighty agency been broken into countless low islets and rocks. The back country is thickly wooded and perfectly flat, a combination which rendered regular triangulation over some portions impracticable. Harbours, to serve as bases of operation, were few in number and difficult to approach. The shore being likewise studded with sunken rocks rising from the bottom very abruptly, made navigation in the vessel while attending on the boats very precarious, but by steaming slowly, keeping a good lookout and continual use of the lead, the vessel only struck one and without any serious result. The broken up character of the shore renders it impossible to measure the number of miles of coastline surveyed.' On one spring voyage in May 1888, his survey ship the Bayfield while on passage to Killarney, Ontario, was 'beset with ice for 96 miles of our passage to Gore Bay, some of which I estimated to have been 20 feet thick caused by one floe being heaved on top of another by the sea, and welded together.' These are conditions more typical of the Arctic.

It was in 1903 that the first hydrographic guide to Georgian Bay *(The Georgian Bay and North Channel Pilot)*, was published by the Department of Marine and Fisheries in Ottawa. It was completed by hand by sextant and lead line well before the introduction of the first gasoline launch, and the gyroscope compass and echo-sounder in the 1920's. Commander Boulton's work in plotting the coastline was a tremendous accomplishment and ultimately made Georgian Bay a safer place. It was only as late as 1964 that the first accurate estimate of the number of islands and islets (100,000) was made by Canadian Coastguard when completing the first detailed charts of the Bay for recreational boaters.

The theme of an unwelcoming wilderness that pervaded Canadian's view of Georgian Bay for much of the 19th century was strengthened by the first geologists to work in the area. As an employee of the fledgling Geological Survey of Canada, Alexander Murray surveyed the Bay and 30,000 Islands for mineral deposits between 1847 and 1858 and remarked that 'a large portion of the region appears to be valueless'. This assessment was confirmed by the Ontario Bureau of Mines in 1899. It wasn't an easy area to work in as Lt. Henry Bayfield had first noted. Here were old hard and highly deformed Precambrian rocks lacking any fossils typical of the uniform younger sedimentary strata that succeed each other in regular order

on the southern and western sides of Georgian Bay that geologists had been trained on in the US and Europe. There was no easy way to date rocks and correlate them from one part of the Shield to another. Travel was obstructed by frequent burnovers left by the lumbermen, a lack of roads and maps, clouds of biting flies and short summers. As late as 1909 the geologist Frank Adams famously spoke of a 'ruined waste' and 'hopeless chaos' when referring to the rocks of the Ottawa-Huron Tract. If it was a difficult place to do geology it was even more difficult to make a living often with tragic consequences.

The Barrens and the roads of broken dreams

A 'barren' is an old Irish word describing a bare limestone pavement; a place where nothing could be grown and thus was of no value. Transplanted to Canada it was used to describe areas of smooth low-relief glacially-scoured crystalline rock along the southern edge of the Shield utterly devoid of soil or sediment. The large trees growing out of cracks in the rock formed a fascinating park-like landscape but gave

Jeft: **Bare limestone plains** *form a wide belt south of the Canadian Shield, across the Bruce Peninsula and Manitoulin Island.*

Above: **Settler's cabin of the late 1860's Parry Sound District** *on Free Grant land.*

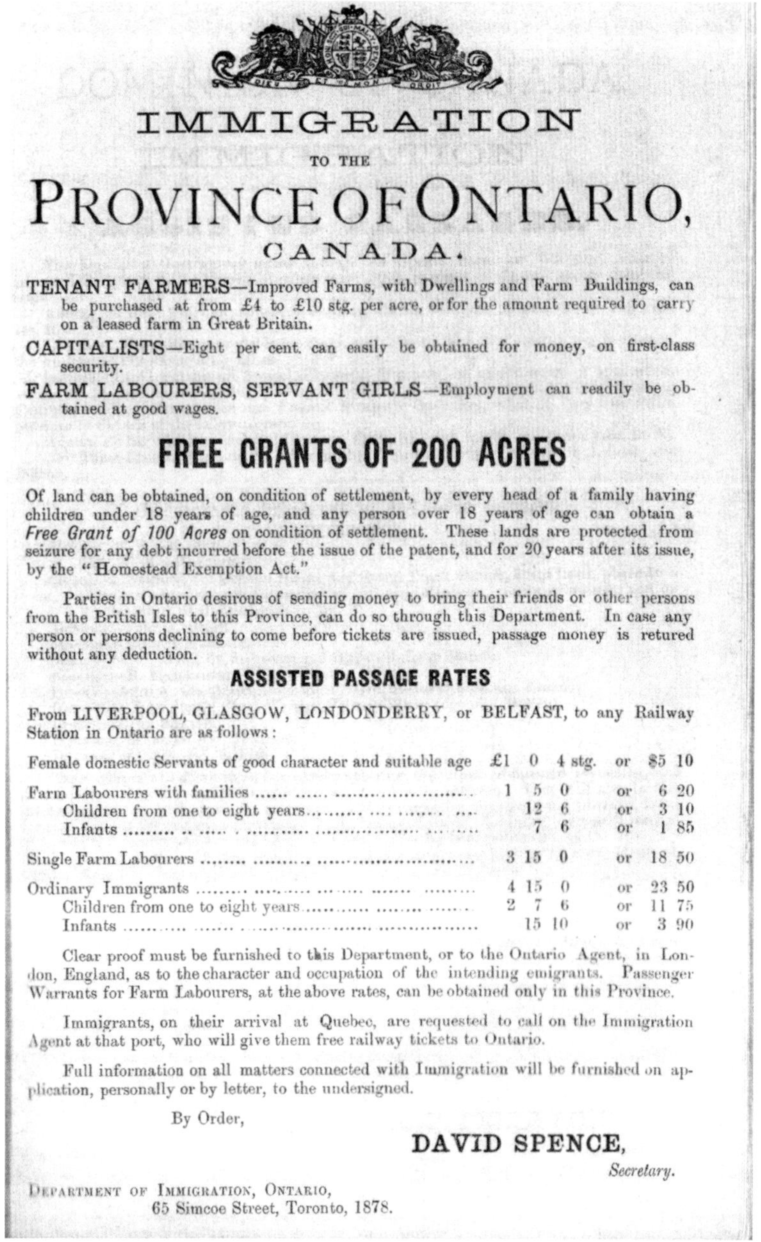

IMMIGRATION

TO THE

PROVINCE OF ONTARIO, CANADA.

TENANT FARMERS—Improved Farms, with Dwellings and Farm Buildings, can be purchased at from £4 to £10 stg. per acre, or for the amount required to carry on a leased farm in Great Britain.

CAPITALISTS—Eight per cent. can easily be obtained for money, on first-class security.

FARM LABOURERS, SERVANT GIRLS—Employment can readily be obtained at good wages.

FREE GRANTS OF 200 ACRES

Of land can be obtained, on condition of settlement, by every head of a family having children under 18 years of age, and any person over 18 years of age can obtain a *Free Grant of 100 Acres* on condition of settlement. These lands are protected from seizure for any debt incurred before the issue of the patent, and for 20 years after its issue, by the "Homestead Exemption Act."

Parties in Ontario desirous of sending money to bring their friends or other persons from the British Isles to this Province, can do so through this Department. In case any person or persons declining to come before tickets are issued, passage money is retured without any deduction.

ASSISTED PASSAGE RATES

From LIVERPOOL, GLASGOW, LONDONDERRY, or BELFAST, to any Railway Station in Ontario are as follows:

Female domestic Servants of good character and suitable age	£1	0	4 stg.	or	$5	10
Farm Labourers with families	1	5	0	or	6	20
Children from one to eight years		12	6	or	3	10
Infants		7	6	or	1	85
Single Farm Labourers	3	15	0	or	18	50
Ordinary Immigrants	4	15	0	or	23	50
Children from one to eight years	2	7	6	or	11	75
Infants		15	10	or	3	90

Clear proof must be furnished to this Department, or to the Ontario Agent, in London, England, as to the character and occupation of the intending emigrants. Passenger Warrants for Farm Labourers, at the above rates, can be obtained only in this Province.

Immigrants, on their arrival at Quebec, are requested to call on the Immigration Agent at that port, who will give them free railway tickets to Ontario.

Full information on all matters connected with Immigration will be furnished on application, personally or by letter, to the undersigned.

By Order,

DAVID SPENCE,
Secretary.

DEPARTMENT OF IMMIGRATION, ONTARIO,
65 Simcoe Street, Toronto, 1878.

Department of Immigration *announcement of 'free grant land' for prospective immigrants from 1878 and the cover of a guide book published in 1879 extolling the virtues of coming to Canada and the Parry Sound District.*

a false impression of fertility. In the mid nineteenth century it was an attractive vista to some since by being 'clothed in hard and soft woods, often of large growth… it gave the promise of good arable soil.' Here then was a new frontier to be exploited by farmers. By 1855, Ontario (called Canada West at the time) was running out of farmland and economic prospects were dim compared to those of the USA, Australia and New Zealand. New immigrants were desperately needed but there was a lack of new land for settlers. It was all about boots on the ground and getting people to live on the Shield then seen as an untapped interior of vital strategic importance if Canada was to grow. It was said at the time that 'the extent of the unsettled country bears so large a proportion to the remaining portions of Upper Canada, that unless settlement can be carried on in this region, Canada would remain a mere frontier strip bordering the margin of the St. Lawrence and the Great Lakes.' Sir William Logan the influential Director of the Geological Survey of Canada declared in 1857 'it appears to me of the highest national importance that the available parts of the Laurentian region generally should be settled upon, as the Devonian and Silurian rocks, hitherto chiefly resorted to, constitute little more

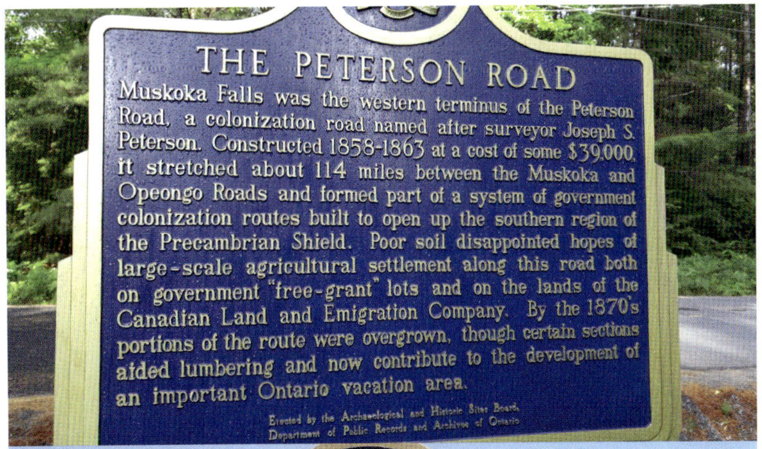

than an extended narrow strip along the frontier of the country.'

The key to future growth it was then hoped would be provided by as many as 20 newly constructed 'colonization roads' that would convey new arrivals from Europe northwards to their new homes. In 1869, the Province passed the Free Grants Act offering land sight unseen on the southern margins of the Canadian Shield designed to attract men 'without capital, for these lands are to be had without price.' Any person over 18 could claim 100 acres and heads of a family with children under the age of 18 could claim 200 acres and could purchase an additional

Colonization roads

For much of the nineteenth century the waters of Georgian Bay were the preferred regional highway from the shipyards and railway terminus at Collingwood with numerous steamers serving the main coastal communities. Inland, by contrast, roads were slow to develop; to the west, the Bruce Peninsula was a cul-de-sac and to the east lay the rocky wilderness of the Shield. The Saugeen Nation ceded the Bruce Peninsula to the Crown in 1854 which resulted in an immediate influx of settlers. The first saw mill built in 1881 in Tobermory accelerated deforestation and with it the inevitable wildfires allowing the peninsula's thin soils to be stripped by wind erosion. The extensive limestone plains that characterise the Bruce today comprise a man-made cultural landscape similar in origin to the to the limestone plains ('alvars') north of Lake Simcoe and eastward to Kingston. The 'Southwest Diagonal' and the 'Central Road' which ran along the backbone of the Bruce were crude waggon trails leading north from Sydenham (now Owen Sound) to Tobermory; these were only to be improved to create Highway 6 as late as 1937.

On the eastern side of Georgian Bay rough 'colonization roads' were constructed northwards from Lake Simcoe and the adjacent Kawartha lakes onto the south flanks of the Canadian Shield under the provisions of the Free Grants and Homestead Act of 1868. These were an ill-fated attempt to lure settlers and farmers north. In reality they were forestry roads intended to open up vast areas between Georgian Bay and the Ottawa River for the lumber industry. It was only in the mid-1950's that Highway 69 from Parry Sound to Sudbury was completed and Georgian Bay was finally encircled by roads. Only one regularly scheduled shipping route survives today; that from Tobermory to South Baymouth on Manitoulin Island.

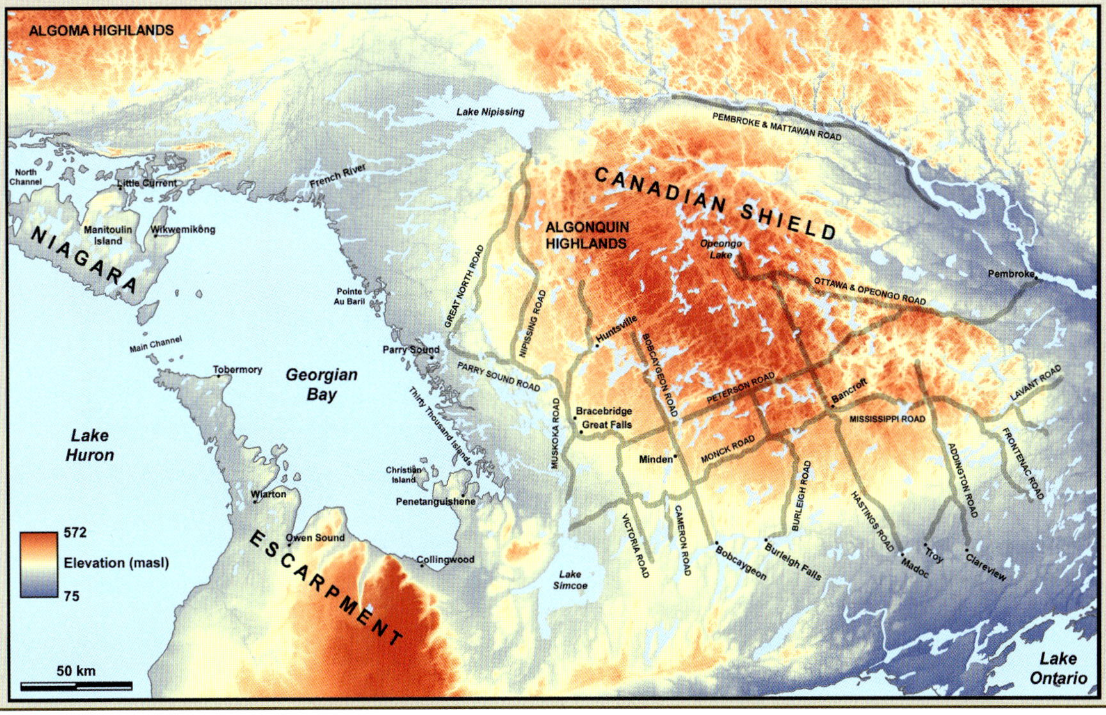

Parry Sound

The harbour filled with timber c. 1910

Stacked cut lumber c. 1915

The tug 'Wave" in Parry Sound Harbour in 1869, the first steam boat used in the harbour. Note mast used for sailing in the event of break downs.

Depot Harbour hockey team in 1920

Parry Sound Trestle Bridge under construction in1905; The longest trestle east of the Rocky Mountains.

John Rudolphus Booth at Depot Harbour

Cutting ice for storage in ice houses c. 1921

Ice sailing in Georgian Bay 1912

All photographs courtesy of West Parry Sound Museum

100. Each settler was given title to their land if they cleared 12 acres in the first year and built an 18 x 20 foot log house.

For a while, the southern Shield became known as 'Last Chance Territory'–a glimmer of hope for many but it soon became obvious that agriculture was impossible on the hard rock barrens. Many families who had dreamt of owning their own farms were duped by overzealous land agents extolling the virtues of free farmsteads in Ontario. Typical of the time are the rosy statements in *Undeveloped lands in Northern and Western Ontario,* an authoritative guidebook published in 1878 for Europeans intending to come to Canada and whose stated aim was 'to encourage settlement and promote immigration: to turn our wild lands into smiling farms and to make the wilderness rejoice and blossom.'

'Ontario has to compete in the Emigration market, not merely with Australia, New Zealand and the United States but also with sister Provinces chiefly with Manitoba which drains us of the younger and more unsettled of our own farming community. Let us therefore lay the claims of our own undeveloped territory of Muskoka and Parry Sound before the public, both in Europe and America so that we may receive our fair quantum of the European swarm and induce the youth of Ontario not rashly to desert their native Province.'

'In one respect nature has been unkind to us… with frowning barriers of inhospitable rocks and many a visitor has recoiled from further entrance to our territory. If he takes Parry Sound with its rocks and yellow sand, or the rock encumbered belt of land which skirts the Georgian Bay, he will probably return by the next steamer to Collingwood.' But, it was noted Georgian Bay and Lake Huron 'possess so extensive a coast on the great waters where American enterprise has so rapidly advanced to great magnitude, the inhabitants will have every money making enterprise of land and water open to them.'

The Editor of the Free Grant Gazette published in Bracebridge went so far as to claim that 'The axe is a pleasant tool to use and conquering the forest is, as it were, conquering an enemy worthy of man's energies.' But, it warned, 'leave the beer barrel and whiskey for ever behind you. If you are strong and healthy as you ought to be,

The lumber freighter 'Simcoe' *being loaded in Parry Sound in 1896. In 1897 twenty three million board feet of lumber were shipped out of Parry Sound; eleven million by the Parry Sound Lumber Company, nine million by the Guelph Lumber Company and three million by the Seguin Mills*

you won't need any stimulants other than the fresh and bracing air of the bush.'

Government reports of the lack of any mineral wealth in the area were roundly rubbished in the rush to populate the Shield. 'Geologists are not infallible' and 'that we may safely doubt that nature, which has enriched the territory to our west in Lake Superior has condemned Muskoka to mineral infertility.' Others even argued that the 'bush hides everything' and once cleared great wealth would be revealed. The discovery of small copper deposits at the Wilcox and McGown sites in 1893 near Parry Sound started a period of active prospecting but nothing of any significance was discovered. As we now know, the rocks of the Parry Sound district were formed at too great depths, with the more shallow mineral-rich strata having been eroded off and carried away by rivers. While the area had no mineral resources of its own it was to some degree affected by the mineral booms elsewhere in northern Ontario, principally Sudbury and the need to ship mineral ores south through Georgian Bay. But this too, was typically short-lived and failed to have any lasting impacts.

Key Harbour is symbolic of the short-lived boom and bust economic development that has characterized Georgian Bay. The harbour facility was built in 1909 at the mouth of the Key River as a transhipment port for iron ore from the Moose Mountain Mine discovered in the late 1880's along the Vermilion River near Sudbury 110 kilometres to the north. Export of ore soon shifted to the much more accessible Depot Harbour near Parry Sound in 1916. Key Harbour had a brief resurgence between 1929 and 1938 when a 500 m long dock was built to handle coal shipped in from Virginia for the Canadian National Railway. The tracks were torn up for scrap in 1960 and though the pilings for the coal dock still survive any evidence of a town site has been lost to the bush.

The mines of northern Ontario were prodigious consumers of explosives and the first dynamite plant opened at Ambo (renamed Nobel) close to but at a safe distance from Parry Sound in 1914. It was greatly expanded during WWI, experienced a brief resurgence in WWII and closed for good in 1984 in which time it was used as a research facility by A.V. Roe for the Arrow/Iroquois jet fighter until cancelled in 1959.

Manitoulin Island (2,766 km^2) is the largest freshwater lake island in the world and politically and culturally considered part of Northern Ontario. Geologically, however it is part of the south, because it is a continuation of the Bruce Peninsula and the Niagara Escarpment. It has the largest northernmost expanse of limestones in Georgian Bay and is separated from the Precambrian Shield rocks to the north by the North Channel. This waterway was used by fur

Left: ***Depot Harbour*** *just after the First World War*

Right: ***Swing Bridge*** *at Little Current joins Manitoulin Island with the mainland built between 1912 and 1913.*

Key Harbour

traders as a protected route westward from the French River as it leads to the St Mary's River at Sault Ste Marie. The island's topography is step-like dominated by numerous abrupt escarpment edges (risers) that face north and wide, prairie-like limestone plains (flats) that have been heavily modified by glacial erosion to form numerous lakes. This unique landscape is however 'at risk' being increasingly threatened by the industrial blight of wind farms. The arrival of the Northern Railroad at Collingwood in 1855 had a knock on effect in northern Georgian Bay as settlers arrived hungry for land. Manitoulin had been settled by many First Nation groups (Odawa, Ojibwe, Potawatomi) after the War of 1812 and by 1830 had become the centre for Aboriginal administration for northern Ontario. It was ceded to the Crown in 1836 who established the island as a refuge for Natives free from European influence. However, in 1866, it was opened up to European settlement and logging (10 years after the Bruce Peninsula) for 50 cents an acre resulting in the re-settlement of its first nation inhabitants into six small Anishinaabe reserves in the northeastern part of the island (now known as the Wikwemikong Unceded Indian Reserve). Many island settlers had initially been drawn to the Bruce Peninsula where initial optimism had been tempered by the realisation that once logged, there was insufficient soil on the limestone plains to grow crops. Logging, and raising turkeys and sheep were the mainstays on Manitoulin.

The railway reached the island in 1913; abandoned in 1980 the prominent swing bridge crossing the narrowest part of the North Channel at Little Current is still in operation for vehicles on Highway 6 from South Baymouth the ferry terminal along the island's south shore, to Espanola and Highway 17 (Trans-Canada Highway) to the north. Highway 6 leaves Manitoulin Island and zig zags up and through the white quartzite ridges of the LaCloche Range, a Precambrian thrust and fold belt. Given the range of rocks on and close to the island it is a favorite training ground for university geology departments and a photographers delight in the fall, a time of red foliage against white rocks.

Manitoulin Island is also noted for its imagery- in art. The Canadian government has designated Wikwemikong as one of several Cultural Capitals of Canada. In particular, the Woodlands School of Art, the brain child of Norval Morrisseau (1932-2007) and Daphne Odjig (b. 1919), considered the grandparents of native art in Canada, bridges ancient Native traditions and imagery with contemporary culture of European and the Americas, to create bold and compelling images.

The Ojibway Hotel *built in 1906 on Ojibway Island at the northern tip of Shawanaga Island near Pointe au Baril is the last remaining example of the grand old hotels on Georgian Bay for city dwellers wanting to experience the wilderness of the Thirty Thousand Islands with all the comforts of home. It closed in 1966 and it now functions as it a summer community centre.*

At one time, Penetanguishene and its excellent harbour was the only European settlement on Georgian Bay and was noted by Governor John Simcoe in 1794 as a potential location for a naval base to control the Upper Great Lakes. Unused in the war of 1812, it was not until 1828, that the British Garrison on Drummond Island then ceded to the US by the Treaty of Ghent in 1814 which restored the border between the two countries to that before the war started, re-located to Penetang. The garrison was accompanied by many French Canadian voyageurs and Métis fur traders who had worked for the British military. In this way, the unique bi-cultural makeup of the area was established and persists to this day. For a time the small community was the main entry point for settlers en route to Georgian Bay. The arrival of the Midland Railway of Canada at Mundy's Bay (re-named as Midland) bypassed Penetanguishene as a port and

Parry Sound in 1910

The same view 2016

the naval base closed in 1856. The town of 'Penetang' is built on the staircase-like terraces left by glacial Lake Algonquin 11,000 years ago.

At one time, southern Georgian Bay and Port McNicoll could claim to be the gateway to western Canada at a time when Canadians travelled westward by train and Great Lakes steamships from Montreal and the Atlantic seaboard. The port is now the permanent home of the last surviving Edwardian-era steamship the SS Keewatin which was built in Govan, Scotland in 1907. She is a passenger liner that formerly carried passengers (including many emigrants new to Canada) from the 'boat train' from Toronto and also freight, between Port McNicoll and Thunder Bay on Lake Superior for the Canadian Pacific Railway's Great Lakes Steamship Service until 1966. Port McNicoll was founded in 1908 when the centre of the CP's operations was moved from Owen Sound.

The port town of Parry Sound was founded in 1857, and its early days were reliant on its forest wealth and the town is forever associated with J.R. Booth (1827-1925) a classic 'lumber baron' who in 1891, completed construction of the Ottawa, Arnprior and Parry Sound Railway (OAPR) from Georgian Bay (Depot Harbour on nearby Parry Island) through what is now Algonquin Park to Ottawa. Depot Harbour was built 10 km south west of Parry Sound partly on land appropriated from the Anishinaabe Indian Reserve (land was deemed too expensive in Parry Sound itself) and was then regarded as the best natural harbour on the Great Lakes and handled lumber, and grain shipped by boat from Port Arthur (Thunder Bay) and then transhipped by rail to Ottawa and Europe. Widening of the Welland Canal to ocean-going vessels in 1932 undermined the need for Depot Harbour which closed for good in 1979 after a series of fires and occasional use exporting cordite and explosive from nearby Nobel in WWII, and iron ore from Sudbury. Once visited by large Great Lakers such as the Edmund Fitzgerald, decaying harbour walls are all that remains today. The famous Rose Point Swing Bridge carried the OAPR onto Parry Island from the mainland; completed in 1912 it still dutifully opens on the hour for passing ships travelling the sheltered South Channel route to the Thirty Thousand Islands and the open waters of Georgian Bay.

James Cleland Hamilton wrote in 1893 of a journey where 'we came to the beautiful village of Parry Sound and along this shore are, on islands and peninsulas, the vacation cottages of families from Toronto and elsewhere, who with sail or steam craft, enjoy in summer the most exhilarating of fresh air at an elevation of 578 feet above sea level.' One of the most famous views of the industrial heritage of Georgian Bay, whether seen from the water, road or rail is of the impressive Canadian Pacific Railway (CPR) Trestle Bridge that crosses the mouth of the Sequin River at Parry Sound. It is 517 m long and was completed in 1907. Today it carries westbound rail traffic for the CPR and the Canadian National Railway and is longest rail trestle east of the Rocky Mountains. Group of Seven associated artist Tom Thompson painted the bridge in July 1914. The work of local artist

Alan Stein perfectly captures the overarching presence of the bridge over the busy colourful port (chapter 12).

Parry Sound had three saw mills in the early 20th century producing sawn lumber for the eastern US and British markets. The difference between the appearance of the town in 1910 'when lumbering was king' and today seen from the high ground where the West Parry Sound Museum is now located, provides a graphic image of the enormous impact of the lumber industry on the landscape. The last mill burnt to the ground in the early 1920's and the lumber industry came to an end.

Located on the southwest margin of Georgian Bay, Owen Sound is the largest community on the Bay and nestled within the narrow steep-sided valley of the Sydenham River that was carved by ice-age glaciers into the face of the Niagara Escarpment, and which continues offshore under the waters of Lake Huron. At one time, when the number of its churches rivalled that of its pubs its busy waterfront once enjoyed a reputation as 'Chicago of the North' benefitting from the arrival of the railway in Collingwood in 1855 and the beginning of the busy steamship route to Chicago. From 1950 to 1963 the SS Norgoma connected Owen Sound with Sault Ste Marie via Killarney, the North Channel and Manitoulin Island in the absence of any improved road that connected western and eastern Ontario until as late as 1960 with final upgrading and completion of Highway 17. The demise of the fishing industry and the prominence of trucking devastated maritime trade. Owen Sound is now the regional hub of Grey County and is strategically located at the base of the Bruce Peninsula and the highway to Tobermory and the car ferry to Manitoulin which still provides a westernmost road route from

One of the most famous views of the industrial heritage of Georgian Bay, whether seen from the water, road or rail is of the impressive Canadian Pacific Railway (CPR) Trestle Bridge at Parry Sound

Baie Fine near Killarney

southern to northern Ontario.

Killarney is an icon hidden in the far northern reaches of Georgian Bay. Located east of the entrance to the North Channel which separates Manitoulin Island from Lake Huron and surrounded by white quartzite ridges of the La Cloche Ranges, rounded glacially-sculpted peaks of pink granite and evergreen White Pines, the Killarney Channel and its small portside community is sheltered from the deeper waters of Georgian Bay by George Island. For ever associated with the work of A.Y. Jackson a prominent member of the Group of Seven artists, it is the oldest community on the north shore of Georgian Bay with a total population of about 500 and lies close to the ghost towns of French River, Collins Inlet and Key Harbour. The site first named Shebahonaning (canoe passage) was settled in 1820 by fur trader Étienne Augustin de Lamorandière and his Anishinaabe wife Josephte Saisaigonokwe. It was only connected by Highway 637 to Hiway 69 in 1962, two years before the founding of Killarney Provincial Park famous for its pristine clear lakes and winding trails. Herbert Fisheries and Fish and Chips on dockside is world famous

Source: Archives Ontario. 1919 Eaton's Toronto Spring and Summer Catalogue; T Eaton Company.

Courtesy of Sears Canada

For more than 100 years, from the 1780's to the 1920's, the inshore vessel of choice among Georgian Bay fishermen was the sturdy Mackinaw sailboat; essentially a large roughly 6 m-long canoe with a flat bottom (for beaching) and centre board allowing a mast and sail. It was popular throughout the Great Lakes because it was stable, fast, light and maneuverable and thus could quickly get fish catches to fishing stations where they could be soaked in brine and preserved. The first outboard motors became available on Georgian Bay after 1914 through retailers such as T. Eaton Company, heralding the decline in popularity of the Mackinaw boat. The 1919 Eaton catalogue lists a 2 HP Evinrude motor for $135 (including freight). Races between competing outboards shown here at Parry Sound in 1927 convey the excitement of the new technology which opened up the Bay for cottagers and recreational fishing and lessened the importance of the supply boats calling from island to island.

and plays the same role (but uses local fish) in inducing boaters and tourists to visit the area that Henry's Fish and Chips does at Sans Souci near the South Channel entrance to Parry Sound. Nearby Baie Fine is a superb example of a deep, narrow glacially-excavated valley now flooded by the waters of Lake Huron.

The French River in northern Georgian Bay was the major link in the fur trade between Montreal, the Mattawa and Ottawa rivers and Lake Superior; in use by voyageurs ('travelers') and canoes as late as 1870, the route now only remembered by lonely portages, place names and songs. The landscape of northern Georgian Bay would have been irrevocably altered had one piece of major industrial infrastructure gone ahead as planned for the French River in the early twentieth century. The 70 km long Georgian Bay Ship Canal was promoted as a project having the potential economic impact of a Panama Canal as it would allow large Great Lake freighters to move directly from Lake Huron to Montreal and thence to the Atlantic. In 1904, the Department of Public Works commenced a detailed field study of the route including its geology and recommended construction of a large lock at the Chaudiere Rapids one of several needed to lift boats upstream to Lake Nipissing at a cost of $100 million and taking ten years to complete. The advent of the railways, especially completion of the Canadian Pacific line in 1882, saw that the canal project was never started; laying down tracks was cheaper and quicker than blasting through hard rock of the Canadian Shield.

On the waters of Georgian Bay 1883 was a big year. The Canadian Hydrographic Service (CHS) was set up under the name of the Georgian

The French River: *Gateway to the upper lakes.*

Bay Survey on August 13 in response to the tragic sinking of the 64 ton, 136 feet long 'Asia' with the loss of 97 people while en route from Collingwood to the French River. The vessel was a flat-bottomed canal boat that had worked on the Welland Canal and was totally unsuited to the rough waters of Georgian Bay. It was also heavily overloaded when it foundered in hurricane force winds and heavy seas northwest of Parry Sound. The CHS's mandate was to survey and map Georgian Bay and Lake Huron and it was subsequently extended to all of Canada's inland waterways and coasts.

The evolution of the fishing industry in the Great Lakes and Georgian Bay has been well told by Margaret Beattie Bogue in her classic 'Fishing the Great Lakes: An Environmental History 1783-1933.' The rise and dramatic fall of fishing as an economic activity in Georgian Bay waters closely parallels that of the logging industry around its landward margins; a largely free resource was there for the taking with no regard for the future or a sustainable industry. The major difference is that while the damaging effects of extensive logging on watersheds were all too easy to see, the consequences of overfishing and the major long-lived structural changes it has brought about to Georgian Bay aquatic ecosystems, are out of sight and largely out of mind.

Atlantic salmon had been virtually wiped out in Lake Ontario by 1860 by destructive overfishing and loss of habitat principally by the damming of rivers for mills. Georgian Bay at that time was considered the most productive of all the Canadian Great Lake waters and fishing employed many thousands of people, indeed it had been a staple of the Fur Trade in the late 1700's. As late as 1830, the Hudson Bay Company had about two dozen trading posts on Manitoulin Island buying fish and fish oil from native peoples. In 1836, the Lieutenant Governor of Upper Canada Sir Francis Bond Head remarked after a visit to Manitoulin Island that the 'waters abound with fish' especially

Left: ***'Few prospects*** *exceed in grandeur and singularity those which are here afforded by groups of long and lofty islets, extending in giant rays from a centre into some dark bay, the clear water reflecting their rugged outlines and wild foliage, and the solemn stillness pervading these solitudes.' J.J. Bigsby's description of the mouth of the French River written in 1824.*

Above: ***Replaced and refurbished*** *on several occasions the Bustard Rocks Range, consisting of three lighthouses, have guarded the entrance to the French River since 1875.*

Mackinaw sailboat; the all-purpose pick-up of choice on the waters of Georgian Bay until superceded by outbaord engines (Courtesy of Collingwood Museum).

sturgeon. The Lake Huron Treaty of 1850 recognized native fisheries but did nothing to protect them from encroachment by non-native fishermen resulting in many conflicts and destruction of fishing stations; even the Fishery Act of 1858 stayed silent on the matter of native fishing rights. The Manitoulin Island Treaty of 1862 gave some protection but was only signed by the Ojibwa people of the western part of Manitoulin Island not the Ottawa people on the Wikwemikong Peninsula to the east. The protection was to prove illusory.

Declines in fish yields in the other Great Lakes resulted in an increasing number of US fishermen in Georgian Bay after 1860 seeking lake trout, yellow pickerel and whitefish. Government legislation after 1860 such as the revised Fishery Act of 1865 essentially yielded Native interests to the commercial fishing industry which rapidly depleted the inshore waters and fish on which local people were dependent. The arrival of the railroad in 1858 in Collingwood was a key event.

Once dominated by numerous individual fishermen using small boats such as the classic Mackinaw, the introduction in the 1880's of the steam tug and the ability to lift long gill nets by steam winches combined with new advances in freezing and storing and transporting fish in the early 1890's, consolidated the industry into a few large US-based commercial concerns. These companies were based primarily in Chicago and Buffalo because of their excellent railroad connections to US markets. Fish rapidly became a staple of the urban working class diet and was widely distributed to mining and lumber camps in the north. Some was even shipped as far as London and Liverpool in England. F.G.M. Fraser, in his annual report as a Fishery Overseer on Georgian Bay reported that 'gangs of fisherman… carry on extensive illegal fishing ….protected by fish dealers who are agents of American firms.' The wholesale dumping of offal produced by gutting of fish into Georgian Bay was also highlighted and condemned. The effluent from sawmills choked rivers with sawdust and wood debris, and catches inevitably declined. Despite the well-meaning work of the Georgian Bay Fisheries Commission established in 1905, and the Federal Department of Marine and Fisheries and its Georgian Bay Biological Station (opened in 1902 but closed in 1914), government much better served the interests of commercial fishery than those of the fish themselves and the communities that had depended on them echoing the use of the forests that surrounded Georgian Bay. The fishery had all but collapsed by the 1930's. The endless tug-of-war between Federal and Provincial control of the fisheries blunted conservation efforts by failing to exert authority over the industry and also failed to protect and reconcile the interests of both non-native communities and recreational fishermen which has still yet to be resolved. Invasive fish species are a new threat. There are lessons here for the future as urban development moves north into Georgian Bay environments that were extensively altered by nineteenth century logging but which have recovered sufficiently to now be regarded as being 'quasi natural.' This recovery has been aided by the strongly seasonal nature of tourism and recreation; the Bay has been protected to a large degree by its long winter and extensive ice cover.

In the end, the wilderness surrounding Georgian Bay never blossomed agriculturally as had been hoped in the nineteenth century; there was simply too much rock and not enough soil. Today, ghost

SS Keewatin at Port McNicoll

The Canadian Pacific Railway *was completed along the eastern shore of Georgian Bay in the early 20th century and railcars brought cottagers north from the major cities of the eastern USA and Toronto to the Thirty Thousand Islands. A mural in Sequin Street in Parry Sound shows a CP railcar of the time with Massausauga rattlesnakes on the rocks.*

communities of rotting broken-down log homes and lonely grassed-over pioneer cemeteries dot the old colonization roads that snake up onto and across the southern Shield. Highway 6 (the 'Chisholm Trail' today but formerly the Monck Colonization Road) starting in the small pioneer settlement of Sebright, the Victoria Road (Highway 36) running north of Balsam Lake, the Parry Sound Colonization Road (Highway 141) that meanders across Muskoka through Rosseau, the Peterson Road through Muskoka, and the Nipissing Trail to the north of Rosseau survive to this day as a testament to a failed vision of 'farming rock.' Truly, these are the roads of broken dreams.

More realistic views of farming the shield began to appear such as that of the *Guidebook and Atlas of the Muskoka and Parry Sound Districts* published in Toronto in 1879 by H. R. Page and Co. 'Now, gentle reader, I must bring you to something on which all the abusive epithets in the English language have been poured out in lavish abundance, a hard, rugged, undeniable fact — the Muskoka rock. It is unfortunate that Muskoka, geologically, does not 'put its best leg foremost' to use the old proverbial advice, but, on the contrary, seems to awe the would-be agriculturist, whether he enter the territory from Parry Sound or from Washago, by showing him the ground-floor of primeval rocks. He stands appalled.'

The most bitter assessment of this unhappy phase of Georgian's Bay history is provided in W.C. Pollard's memoir of Life on the Frontier (1910) where he describes the opportunities created by completion of the transcontinental Canadian Pacific Railway (1885) to people who were discovering that the lands they had cleared so laboriously on the Shield were worth little. 'When the majority realized just where they were and the prospects ahead, they marveled that they had ever attempted to remain among the rocks of the Parry Sound District.'

Instead of a farmer's paradise, the Shield became 'Great Britain's wood yard' and the source of massive, squared pine timbers for its rapidly growing cities and immense shipping fleets. Special docks were built in Britain to handle Canadian lumber. The forests were literally mined and huge tracts were devastated by cutting and swept by bush fires underscoring the harsh rocky face of the land. A constant theme that echoes around the turn of the century after the abrupt collapse of the lumber industry in the 1890's is that land originally cleared for farms would be more valuable if it were reforested. W.C. Pollard stated 'What economic loss the destruction of these forests meant to the Dominion of Canada, only time will tell, as most of the lands after they were cleared were unsuited for agricultural purposes and should never have been thrown open for settlement.' In the first modern survey of the geology of the area (which confirmed the absence of any mineral wealth) Jack Satterly (1942) considered that many farms were 'useless land' and ought to be repossessed and reforested. In fact, that is what ensued unintentionally. Postcards and photographs from the turn of the century depict communities surrounded by bare sterile rocks surgically cleared of any tree cover, one is forcibly struck by how much the forest has recovered since.

Many towns amidst the rocky expanse of the surrounding Shield in Central Ontario especially in Muskoka such as Haliburton, Huntsville, Gravenhurst, Bracebridge and many other smaller communities such as Rousseau and Raymond all got their start as farming communities working glaciolacustrine sediments left from the last ice age, especially those left by glacial Lake Algonquin. In northern Ontario, the enormous 'Clay Belts' near Kirkland Lake, Cochrane and New Liskeard are of similar age and origin and were vigorously promoted in Europe as 'new lands' in the late nineteenth

***The Norisle** was built in Collingwood in 1946 and ran until 1974 in the summer months between Tobermory and South Baymouth on Manitoulin Island . Now at the Manitowaning Heritage Centre on Manitoulin Island there are plans to sink her off Tobermory as a dive site.*

and early twentieth centuries but their cold wet soils and short growing season prevented much development. Today, the success of the 'glacial lake oasis' communities of Muskoka obscures their glacial origins and ironically mirrors the extensive use of the shoreline of glacial lake Algonquin by Paleo-Indian peoples some 12,000 years ago.

The rocks get tamed and a new breed of settler moves in

By the close of the 19th century the fruitless task of farming the Shield of the Parry Sound District didn't really much matter. The recalcitrant rocks were now being tamed with dynamite being manufactured ironically, in the heart of Georgian Bay at Nobel, near Parry Sound. Now the rocks could be carved like butter for roads and railways. By the end of the nineteenth century Ontario was also changing and farmland was no longer needed to attract a new class of emigrants; the Province was becoming an urban economy centred on manufacturing. Mining was taking off in the north where successive mining booms notably at Sudbury and Cobalt were creating well paid jobs.

The Yukon Gold Rush of 1898 had drained the Shield of so many people that the Parry Sound Directory for 1898-9 felt it necessary once again, to herald 'its immense tracts of arable land, vast areas of mineral wealth and excellent system of transportation'; all fabrications echoing the rampant earlier boosterism surrounding the Free Grant Act of 1869. But it also noted in an altogether much more sober and accurate harbinger of the future that 'its many attractions to the tourist in search of scenery or sport conspire to make trips both pleasurable and healthful'; the absence of hay fever was regarded as being of especial note.

By the early twentieth century, colonization roads were no longer needed and many fell into disuse. The southern Shield and Georgian Bay now lay within easy reach of Toronto courtesy of new railroads and coastal steamers. What was previously regarded as a 'worthless tract' was now being considered invaluable for recreation. Tourism was spurred by the artwork of the Group of Seven who successfully transformed wilderness into a national icon for the growing number of urban dwellers with time on their hands seeking solace in the backwoods after the horrors of the First World War. The axe would rule no more.

By the late 1930's the new breed of seasonal settler was being carried comfortably northwards by automobiles riding on newly-built tarmac roads constructed as make-work projects in the Depression years. Taking advantage of new technology such as outboard motors, they would re-colonize the forests and islands of Georgian Bay.

Today, hunting and gathering on the Shield still continues in the form of finding new mineral resources from its rocks and the basic north-south contrast still survives intact and is sharpened with every passing year. Old, northern Canada is seemingly content to draw on its deep roots in the landscape and ancient rocks whereas the new urban Canada in the south is forever seeking to reinvent itself on an ever-expanding built landscape, and in the process becoming no different than anywhere else. But that is where most of us live and here still, is the source of excitement in driving north and crossing the southern boundary of the Canadian Shield, seeing the familiar world of farms and towns slowly disappear and those old rocks and a strange new rockscape emerges.

Ironically, by preventing farming and urban development, the inhospitable rocky barrens of Georgian Bay preserved large areas of natural habitat close to Canada's largest urban centre.

White pine lumber being stacked onto wooden sailing vessels along the wharf of the Victoria Harbour Lumber Company c.1890. (Courtesy of Huronia Museum Post Card Coll., Midland.)

CHAPTER

10

THE EVOLVING GEORGIAN BAY FORESTS

JAMIE HUNTER

Rafts of logs *being stored until ready for the mill at Victoria Harbour Lumber Company c.1900. (Courtesy of Huronia Museum Post Card Coll., Midland.)*

The vital role of forests in human cultures: a 10,000-year-long history

Native peoples in the Georgian Bay region have extensively exploited forest resources since the boreal forests re-colonized the area toward the end of the last glaciation some 10,000 year ago and the final drainage of glacial Lake Algonquin. For subsequent millennia many First Nations groups thrived on the forest and used its products to enhance and sustain their way of life. They skillfully crafted tools, homes, and unique transportation devices and they extensively cleared forests for agricultural fields, and traded wooden products throughout eastern North America. So important was the forest that the Georgian Bay First Nations believed that in death a voyage to the land of the souls was made that required travelling along a trail to the west, through a dense forest, and crossing a giant log that bridged a river on the road to the afterlife.

For the Georgian Bay First Nations forest products were a vital part of life, a resource to be exploited and used to the fullest. In 1615 when Samuel de Champlain arrived in Wendake, home of the Huron/Wendat people living in the Georgian Bay region, he remarked on how wonderful this country was with its open meadows and farmland and its numerous villages. Champlain quickly realized that the thousands of residents, whose lives depended on the heavily forested areas of Georgian Bay, would want a lasting alliance with the "iron men"– traders of metal cutting and chopping tools. These tools would ease the labour in working with wood and make possible techniques still unknown to the First Nations.

Similarly the 17th century French missionaries to Georgian Bay constructed tools, homes and transportation devices from the forest. They built the first European vessel to sail on the Upper Great Lakes. Called a shallop, it was constructed at Sainte-Marie among the Hurons (near present-day Midland) in 1643 and was at least thirty feet in length. Relying on the forests, the French fur traders cleared lands for their commercial and domestic needs and grew crops, built communities, and in the 1670s constructed the Griffon – a wooden sailing vessel also on the Upper Great Lakes. This reliance on the forest continued and by the early 1800s the construction and repair of ships from forest materials, including planking, knees, keels and ribs, kept the shipwrights and artificers of the Royal Navy employed at the Schooner Town (present-day Wasaga Beach) and Penetanguishene depots.

By 1830 the first stand-alone sawmill was built by the Indian Department of the British Military for the Ojibway people of southeastern Georgian Bay. The British government had established a reserve at Coldwater for the Ojibway and a requirement of reserve living was to build your own house. A dedicated sawmill produced the boards used in the construction process. Soon after, the squared log industry developed throughout Ontario, including Georgian Bay, and logs were shipped to Quebec City for transshipment to European destinations. In the 1850s entrepreneurs constructed small water driven mills to replace the labour intensive hand sawing techniques. These mills were superseded by massive steam-fired milling operations.

The Ontario forest product industry up to the mid 19th century had for the most part been based on red and white pine and as the rich forests in the Ottawa Valley were depleted the industry moved into the Georgian Bay basin. American lumber merchants from New York and

Pennsylvania bought the smaller Georgian Bay sawmill companies and the industry boomed as British and American capitalists invested in larger volume milling operations. The extraction, towing, shipment-to-markets as well as the wholesale and retail selling of forest products took on enormous proportions. Provincial Crown lands of red and white pine were leased to forestry operators and from 1870 to 1900 the red and white pine industry generated over 30% of the annual government revenues for Ontario and 20% from 1900 to 1920.

Not surprisingly the continuous commercial exploitation of pine forests began to exhaust the resource to the point where most large operators were forced to shut down. Seeking new opportunities to supplement the shrinking pine industry, the smaller milling operators looked to the forest for other options and they found diversification in the use of hemlock for the animal hide tanning industry. Subsequent to this, logging a variety of the hard woods to meet new consumer interests served to sustain forest productivity albeit significantly reduced. At the end of the 1890s the pulp and paper industry had moved into the forests of Ontario and Georgian Bay and established itself as a lucrative business.

By the early 1950s the nature of the forest industry had changed dramatically. The large operators that had managed to maintain their operations in spite of declining pine resources ceased operation because of the technological advancements in the industry. Gas-fired chain saws, mechanized skidders, diesel trucks to haul away logs, and very sophisticated milling processes, no longer demanded the large work crews of the past to cut, haul and transport the resource. Today the forest industry is still important in the Georgian Bay basin, but it is nothing like what it was in the heyday of the 1880-1910 logging era.

Native uses of the forest

All Georgian Bay First Nations relied on forest products. The archaeological record and early documents written by Europeans reveal a dependency on the forest for tools, ornaments, leisure items, elaborate houses and various forms of transportation devices including snowshoes, toboggans, sleighs, dugouts and birch bark canoes. This wooden technology changed dramatically over millennia and became vital in the complicated long distance trading relationships that developed between the Iroquoian groups including the Wendat and the Petun of Georgian Bay, with others in eastern North America.

Birch bark canoe, *12 to 14 feet long, under construction showing cedar ribs, sheets of birch bark, spruce root and pine pitch. (Courtesy of Sainte-Marie among the Hurons Midland.)*

***A long distance birch bark** "trading canoe" brought a cannon, cows, pigs, chickens, and all types of supplies to the St. Joseph Mission, Jesuit Headquarters, at Sainte-Marie among the Hurons 1639-1649. (Courtesy of Sainte-Marie among the Hurons Midland.)*

Specialty woods and manufactured objects were highly sought and while commodities exchanged hands over vast distances, long-lasting inter-tribal relationships were formed.

Although still dependent on aquatic resources and the use of seines, nets, and weirs used in fishing, the introduction of agriculture about 1,000 years ago saw the Iroquoian and Algonkian populations clearing sizeable areas of land to cultivate corn, beans, squash, sunflowers and tobacco. As these groups increased from small, localized river valley populations to regional and ethnically distinct nations, each occupied their own defined space. An agricultural way of life resulted in villages increasing in size from up to ten long houses to up to one hundred long houses. These were sophisticated permanent dwellings that incorporated specialized features like defensive mechanisms that kept out the elements, unwanted wildlife like wolves and bears, and the marauding enemy.

As villages grew so too did the need to obtain poles, bark, and firewood to sustain the expanding population. This put a greater strain on local resources as thousands of twenty-five to forty-five foot poles and thousands of linear feet of bark, usually elm, were harvested for constructing longhouses and defensive palisades. Archaeological excavations of the early historic Wendat Village, known today as the Ball Site, AD 1580-1600 uncovered evidence of tens of thousands of longhouse and palisade poles that would have been harvested from a catchment area estimated to be 5,000 acres. The significance of forest products in the evolution of permanent sedentary villages for the First Nations people of Georgian Bay cannot be overstated.

Forest resources beyond settlement and defensive purposes were fundamental for the Georgian Bay First Nations people. The forests were rich in the symbolic, mystical and spiritual meaning they held. For many Iroquoian groups the white pine tree was symbolic of the four cardinal directions and it was the home to the eagle and other important birds. The forest held curative and pharmacological products, like the inner bark of the willow tree, that were used to treat diseases. Boiling white cedar bark as a way to prevent scurvy was well known to First Nations and was administered to the first Europeans when they arrived in northeastern North America suffering from the disease.

Exploiting the forest continued with the arrival of European traders who introduced a wide assortment of metal tools like knives, awls, axes, punches, drills and saws. These new implements enabled First Nations to craft a greater range of objects using new woodworking techniques. Production of utilitarian and symbolic wooden objects including bowls, spoons and masks flourished.

Arrival of the French on "La Mer Douce" – Georgian Bay

The French 'truchement', or company man, Etienne Brule arrived in Wendake in 1610. Recognizing the potential bounty of the fur trade he quickly cultivated relationships with the skilled Wendat traders to ensure that they would only exchange their furs and other products with the trading company led by Samuel de Champlain. This trade was evident at the annual 'tabagies' or trading fairs held for two to three weeks in July just below the La Chine Rapids in Quebec. During this time the western 'Hurronnes' (anyone living west of the Ottawa River) would come down from the 'pays d'en haut' (the upper country of the Canadian Shield) to trade with the French. In 1614 the Wendat and French established a formal trading alliance that obligated the Wendat to trade only with the fur trading monopolies managed by Champlain. As part of the agreement the French were permitted to send missionaries to live among the Wendat people. The agreement also stipulated that the French would support the Wendat in their long-standing warfare with the Haudenosaunee (the Iroquois League) in Upper State New York.

Champlain and his men, and the Recollet priest Father Joseph Le Caron, arrived in Wendake in early August of 1615 within days of each other. The French initially lived within the palisaded Wendat villages and with the Wendat people in their longhouses. Soon after they built their own longhouses and modified the traditional open concept to include defined spaces including a chapel, a private living area and a public area – this type of partitioning was a real novelty for the Wendat.

Missionary work continued in Wendake and when the Recollets lacked the personnel and financial resources necessary to expand the mission they invited the Jesuit religious order to support their cause. Under the continued guidance of Jesuit priest Jean de Brebeuf, Superior of the new Huron Mission, two chapels of European architectural design were built in 1637 and 1638. Made completely from wood they each had a bell tower and brass bell to summon worshippers

***Small sailing craft** similar to a shallop, the first European vessel built on the Great Lakes in 1643 for moving building supplies and people around the Sainte-Marie Mission. (Courtesy of Huronia Museum Midland.)*

The recreated Sainte-Marie among the Hurons, *the first wooden European community on what is now Georgian Bay. (Courtesy of Sainte-Marie among the Hurons Midland.)*

living in the villages of Ossossane and Teaunaustaye. Some villages also adapted European defensive design elements learned from the Jesuits. This was most often reflected in the palisade walls of villages having square corners, each with a firing platform. The introduction of metal saws, hammers, nails and hinges, all tools the Wendat did not have prior to this period, resulted in the building techniques of longhouses and all items made from wood to change in a way that would never be reversed.

In 1639 the new Superior of the Huron Mission, Father Jerome Lalemant, constructed a central Jesuit mission headquarters in Wendake to support the growing efforts of the priests. Paid French workers (engages) hired by the Jesuits, French volunteers, known as donnes, and Wendat volunteers cleared the land and built a palisaded European-styled community completely by hand. Wall and floorboards were sawn, beams were squared, and doors, sills and furniture were manufactured. All were resourced from the Georgian Bay forests. Over the next ten years the headquarters, called Sainte-Marie among the Hurons, thrived and approximately twenty structures were built under the supervision of master builder Charles Boivin. The French style shallop (small sail boat) transported building materials to the river-front mission and ferried people across the river to the extensive crop area that provided agricultural self-sufficiency to the mission. By 1648 Sainte-Marie among the Hurons was the home of nearly seventy French priests, lay brothers, engages and donnes. In short this small community had all of the visual elements of a town in France, European in all ways. There is no doubt that by this time the metal implements of the 'iron men' had made their way to all of the surrounding villages of the Wendat, Petun and Ojibway to the north and to the southernmost parts of Georgian Bay, and also to the villages of the Neutral and the Haudenosaunee further to the south. These villages were within the reach of the missionaries and each had felt the impact of the new technology in all aspects of working with wood.

The Haudenosaunee waged war on the Wendat Nation throughout this period. In 1649 they forced the French and Wendat to abandon Sainte-Marie among the Hurons. Seeking refuge on Ahouêdoë Island (Christian Island in Georgian Bay) the French and Christian Wendat built Sainte-Marie ll. A large raft crafted from squared timbers and sawn lumber, salvaged from the buildings at Sainte-Marie, was sailed to the new location to be used in the new construction. This time a fortified stone building was built to house and protect the French and the Wendat. Tragically, crop failure, famine, and continued warfare with the Haudenosaunee forced the French to return to Quebec accompanied by a small number of Christian Wendat in 1650, while other Wendat took shelter among the Petun, Neutral and Ojibway people to the south, west and north. In time the Iroquois dispersed these groups and they travelled as far west as Red Wing Island in the Mississippi River and then to Ignace Michigan, Manitoulin Island, and the Detroit Windsor Region by 1701. In each instance the people created villages with churches and modified longhouses all built upon new French designs and construction techniques.

In the latter part of the 17th century fur trader and missionary activities pushed further west to Lakes Michigan and the Mississippi River Basin where French styled wooden homes, agricultural plantations and wooden sailing vessels, formed a Native trading empire that travelled along the fur trade corridors of Georgian Bay and the Great Lakes basin. For the French fur traders and the missionaries forest products had been pivotal to their success.

The recreated Naval Dockyard at Penetanguishene representing the period 1817-1834. (Courtesy of Discovery Harbour, Penetanguishene).

The British Military and the forests

The visionary for a naval arsenal on Georgian Bay was the Lieutenant Governor of Upper Canada, John Graves Simcoe. He assumed his new duties in 1791 and one of his first orders of business was to move the capital of Upper Canada from Newark (Niagara on the Lake) to York (Toronto).

In the fall of 1793 Simcoe set out north to visit Lake Huron and during this trip he identified Penetanguishene Harbour, on Georgian Bay, as the perfect site for a naval and military establishment. Its strategic location could effectively position the military to protect the new capital from any kind of overland attack launched from the north. Simcoe instructed Deputy Surveyor Alexander Aitkens to survey the land and the parcel became known as the Penetanguishene Purchase. The negotiations for the purchase were concluded in May of 1795 between Simcoe and Ojibway Chief Yellowhead and were finalized by

Above left: **Painting depicting Lieutenant Henry Bayfield's survey** *of Lake Huron/Georgian Bay in the wooden gigs Ramsden and Troughton among the islands near Moose Point, 1818-1822. (Courtesy of Peter Rindlisbacher, Marine Artist.)*

Above right: **Sawpit at Naval Dockyard**, *Penetanguishene, where logs were cut using a whip saw and two sawyers, one above and one below. (Courtesy of Discovery Harbour Penetanguishene.)*

the Legislative Assembly of Upper Canada in May of 1798. Although now owned by the British the development of naval and military depots on Penetanguishene Harbour did not occur for some time.

It took the War of 1812 between the British and the Americans to spur development on Georgian Bay. The deciding factor was the defeat of Captain Robert Barclay and the British Royal Navy at Put-in-Bay on Lake Erie, by the Americans under the command of Oliver Hazard Perry, in the fall of 1813. The British faced escalating pressure from the North West Company fur traders to provide protection for them from the Americans in their travels through the upper Great Lakes. An overland route from Penetanguishene would allow the Northwest Company to transport its furs from Lake Huron to York without having to use the circuitous and dangerous route through Lake Erie along the American frontier. A second factor was that Penetanguishene was on the route that transported vital supplies from York to the British troops at Fort Michilimackinac, in northern Lake Huron, and Britain's First Nations allies who lived as far west as the Mississippi River.

To this end naval depots were constructed at Schooner Town (present day Wasaga Beach) in 1813 and Penetanguishene in 1814 to provide a presence at Upper Canada's 'backdoor'. They would help to protect the fur trading interests of the Northwest Company. In addition they would store supplies that were destined for the British military troops at Fort Michilimackinac and their western First Nations allies during and after the War of 1812. A flurry of construction activity occurred at both locations as storage buildings, residences and trades shops were built. At Schooner Town three wooden transport ships, *Bee, Wasp* and *Mosquito* as well as numerous other small wooden vessels were constructed.

After the War of 1812 there was no immediate need to protect Upper Canada from a northern attack and Schooner Town and Penetanguishene were ordered closed. But the long, narrow and deep harbour of Penetanguishene Bay was unparalleled on Georgian Bay and its overland access to York via the Penetanguishene Road caused the British to re-establish its presence. The naval establishment at Penetanguishene was redefined in 1817 and it became the navy's main depot on Georgian Bay as well as the navy's new centre for Naval Hydrographic studies. Having done so poorly during the War of 1812, in part due to not being able to navigate through the rocky waters of the Great Lakes, the British dispatched Lieutenant Henry Bayfield to chart Lake Huron using Penetanguishene as his headquarters. The depot also remained pivotal in supplying the military troops and the First Nations allies still living in the north.

Again a flurry of activity ensued at the revitalized Penetanguishene Naval Establishment. An enormous storehouse was constructed to hold necessary supplies that were sent to the north. Land was cleared, and houses, shops, and a hospital were built all to service the growing community. The primary role of the naval and civilian personnel at Penetanguishene was to maintain the two ships H.M.S. *Tecumseth* and H.M.S. *Newash* in a state of readiness should hostilities with the Americans re-emerge. As part of the Rush-Bagot Agreement between the British and Americans both sides had to reduce the number of ships they sailed on the Great Lakes. The British strategy was to maintain two large ships at Penetanguishene and keep them in a state of readiness should hostilities with the Americans re-emerge. Civilian artificers and shipwrights were employed to carryout this massive undertaking. They turned to the forests of Georgian Bay for all the wooden materials they needed for ship repairing and for replacing masts, spars, planking, decking, knees and blocks. This was no small accomplishment at a remote and developing outpost. The Quarterman of Shipwrights at the Penetanguishene Naval Establishment made requests for the installation of a water-powered sawmill but his requests appear to have fallen on deaf ears. There was little doubt that the volume of the various forms of sawn wood needed to meet the building construction needs of the growing population and the ship repair needs of the artificers and shipwrights could never be met by the traditional pit saw and whip saw technologies that were current at the time.

The British Army's Indian Department

Further impact to the forests came from the agreements made between the British and the Americans after the War of 1812 including a boundary commission to establish the border between the United States and British North America territories. The outcome was the ceding of Drummond Island, the location of the British Military Establishment on Lake Huron, to the Americans. The ceding of Drummond Island meant that the British soldiers, fur traders, First Nations allies and Métis personnel associated with the fort had to be relocated. Some went to Amherstburg. Others joined the personnel at the Penetanguishene Naval Establishment and they formed a new Military Establishment. Eventually the First Nations who came with the military personnel established themselves at what became the community of Coldwater, about 30 kilometres east of Penetanguishene.

Again the resources from the forests were required to support all of the building and construction efforts that resulted from the development at Penetanguishene and Coldwater. The issue of a sawmill was raised again as the pit saw and whip saw technologies could not meet the construction demands. This was a reasonable request considering that General Hare of the British Commissariat Department had already noted in 1828 that in the absence of a sawmill at Penetanguishene, lumber would have to be hand sawn or shipped from New Market, a distance of about 76 miles. Both were not cost effective options.

W.H. Bartlett *engraving showing log rafts descending the St. Lawrence River to Quebec City for transshipment to Europe 1842. (Courtesy of Canadian Scenery 1842.)*

It was in Coldwater that Captain Thomas Anderson, of the British Military's Indian Department, established the first reserve for First Nations in Upper Canada under the authority of Lieutenant Governor Sir John Colborne. A condition of reserve living was that each family was required to build their own home. The Indian Department agreed to build a mill at Severn Mills (Port Severn) where there was ample water power to operate it, and there were available and accessible logs from the Severn watershed located above the mill. The sawn products were easily shipped by boat from the mill to Coldwater and to Penetanguishene. As technology advanced there were efficiencies in the lumbering process and in water powered saws that resulted in increased speed in which logs could be milled. The benefits of the mill at Severn Mills were twofold: it provided the First Nations with the construction materials they needed for their homes; and it employed the First Nations living at Coldwater. Unfortunately the mill was located quite some distance from the reserve and while the potential for this operation was promising the distance to the mill proved problematic for the people. It operated for only a short time.

Expansion of Water Powered Mills

All along the southern and southwestern shores of Georgian Bay many small water powered sawmills developed between 1830 and 1860. These mills produced a range of products usually for the immediate regions where the mills were located. In general these mills contained a single upright gang saw and they turned out logs, beams, lumber, lathe and barrel staves from logs that were purchased mainly by settlers.

Throughout this period lumbering continued to evolve and expand. British settlement polices allowed farmers to buy land from the Crown but usually not until after the valuable timber had been removed by lumbering companies and a fee had been paid to the government. In 1850 the British Government negotiated the Robinson-Huron Treaty with the Ojibway First Nation concerning land and island ownership, fishing rights, mineral rights and timber rights over the vast area of Canadian Shield east and north of Georgian Bay that stretched from the Severn River north to Sault Ste. Marie and beyond. This treaty meant that all of the resources once owned by the Ojibway in this geographical area were now the property of the Crown and it could sell all of the various rights associated with the land.

Another important development occurred in 1855 when the Ontario Huron and Simcoe Railway reached Collingwood from Toronto. This expansion created a regional building boom throughout Georgian Bay as the area opened up and people moved in to settle. There was a corresponding need for all forest products for new building construction and for the production of thousands of railway ties.

At the same time a new free trade agreement was negotiated between Canada and the United States for all natural products. Known as the Reciprocity Treaty of 1854 it opened Canadian forest products to the American market. Many of the smaller mills were now expanded to produce export products not so much to Britain as had the square timber trade, but now a variety of milled lumber products could be sold directly to buyers in the United States. Technological innovation continued to improve and the new circular and circular gang (multiple) saws enabled mills to produce significantly greater volumes of lumber than ever before. Added to this was the much greater availability of horses that slowly replaced the less efficient oxen as beasts of burden.

Lumber companies bought thousands of acres of standing timber limits from the Crown. The Government of Upper Canada continued to own all forests even those on private property and it

Left: ***Photo of Playfair Mill 1887-1914*** *formerly the H.H. Cook Mill 1871-1886 showing ladder, mill and stacking piles/railyards in Midland. (Courtesy of Huronia Museum Post Card Coll., Midland.)*

Above: ***Aerial view of Georgian Bay Lumber Company Mills*** *at Waubaushene 1872-1942. (Courtesy of Huronia Museum Post Card Coll., Midland.)*

Interior of milling operations at Georgian Bay Lumber Company Mill at Port Severn c.1890.

had the lands surveyed and divided into townships. But the process for paying the government for timber on its Crown lands was poorly managed and large-scale sales needed to be regulated. To help solve this problem timber on Crown lands, after 1827, was sold by license at public auction to the highest bidder. Following the Act of Union of the Canada's in 1841 which re-united Upper and Lower Canada the government passed the Crown Timber Act of 1849 which set out a framework for the sale of timber on Crown lands. Some of the regulations remain more or less in place to this day.

The Square Timber Industry

Evidence for an extensive square timber business throughout Georgian Bay is scant. George Caswell (1812-1885) immigrated to Canada from Ireland in the early 1830s and went into the lumber business. He was the first lumberman to ship square timber from Georgian Bay to Montreal and from there overseas to England. Lumbermen using teams of oxen hauled logs to rivers or shoreline areas where the logs were squared by hand and then cut into sixteen-foot lengths. All of the square timber was assembled into rafts and floated to Quebec City where the rafts were dismantled. The square timber was then loaded onto larger ocean going vessels and shipped to the European market.

The sheer distance to get square timber from Georgian Bay to Quebec City before the days of the railways was formidable. In the early 1850s two large sea going ships, the *Reindeer* and the *Sardinia,* were built at Matchedash Bay near Coldwater. They were loaded with squared oak timber and shipped to England where the cargo of timber and the vessels themselves were profitably sold. This was the only lumbering initiative of this kind to be recorded in the Georgian Bay basin.

The Georgian Bay square timber industry, for export to European markets, was never of the scale that it was in the Ottawa River and Lake Ontario regions. Georgian Bay lumber entrepreneurs were soon to learn that more money could be made by just milling and not squaring the logs at water driven mill sites, and then shipping them by vessel to the growing American markets.

Fishing and logging

Commercial fishing had started in Georgian Bay as early as 1830 and this increased proportionally with the decline of the fur trade after 1850. In 1868 some 450 men were employed in the Bay and landed over 1,200,000 lbs of whitefish, then 700,000 lbs of trout and 8,000 lbs of pickerel. The peak of the industry was the last two decades of the nineteenth century when 3,229,000 lbs of whitefish were landed in 1888-9 mostly for the fish markets of Buffalo and Chicago. The industry was recognized as one of the pre-eminent fisheries in all of Canada and busy seasonal fishing camps were set up in Georgian Bay on the Mink, Bustard and Bass islands in harmony with the annual cycle of fish spawning along the rocky shorelines of the sheltered inlets in the fall and their migration out to deep cold water of the open Bay in summer. By then, there were conflicts with the lumber industry who in the process of clear cutting entire watersheds dumped huge volumes of waste into rivers and inlets and whose log booms swept away nets or clogged them with woody debris. It was widely observed that as the logging proceeded so the fish disappeared as their inshore habitat was profoundly altered. Clearly it was not sustainable and the fishery's final collapse in the early 1950s is reminiscent of the collapse of the east coast Grand Banks cod fishery in the 1980s after decades of overfishing and lack of any strategic management.

Georgian Bay Mills Expand

As related in the previous chapter throughout the 1860s and 1870s Southern Ontario could no longer accommodate the influx of immigrants with agricultural land and they began to move north on to the Barren Lands of the Canadian Shield surrounding Georgian Bay. The Crown Lands Department had changed its policies and all immigrants were now required to settle land with agricultural productivity being the outcome. Much of the land was totally unsuitable for agriculture. Unable to make a living as farmers they turned to logging or working in the sawmills to augment their meager incomes from their less than successful crops.

In 1868 the Ontario Free Grants and Homestead Act was passed, which made land available to settlers at no cost throughout the eastern shore of Georgian Bay and the interior lands east to the Haliburton Highlands. From approximately 1870 to 1930 an assault on the Georgian Bay forest was carried out on a massive scale. Free land saw the expansion of the small communities located in the vicinity of the water driven mill sites that had been developed along the rivers and river mouths. The scale of lumber operations and mills expanded dramatically. Further to this, population growth was promoted by many of these communities becoming destination stops for the railways including the Northern Railway to Owen Sound in 1869 and Penetanguishene in 1878, and the Midland Railway to Waubaushene in 1876, and to Midland in 1879.

Pulp and Paper Forest Products 1890 to the Present

Logging and milling defined the primary forestry industry on Georgian Bay until almost the end of the 19th century. As the century was nearing its end another business opportunity was quickly developing that would have a profound effect on the forests – the pulp and paper industry. Most of it would be located in northern Ontario west of Sault Ste. Marie, but a sizable portion would be located from Sault Ste. Marie east to the Spanish River in the Georgian Bay basin.

Lumber stacks for the Letherby and Chew Lumber Company at Midland with Aberdeen small grain elevator and Grand Trunk Railway in background c.1906. (Courtesy of Huronia Museum Post Card Coll., Midland.)

Historically paper had been manufactured from rags. The process to make paper from wood fibres treated with sulfuric acid was developed in 1867 in the United States. White and black spruce varieties were the preferred tree species but others like poplar, balsam and fir were also suitable. The Spanish River Pulp and Paper Company emerged and developed mills located on rivers that drained into northern Georgian Bay where vital hydroelectric power supplies were available. It became the primary pulp and paper producer within the Georgian Bay basin.

The pulp and paper industry grew rapidly and American supplies of spruce could not keep-up with the demand for newsprint. This was complicated further by the 1891 McKinley Tariff Act that had imposed a 15% duty on all Canadian pulp and paper products imported into the United States. This included everything from sawn logs to paper. This tariff significantly reduced American investment in Ontario forest products.

By 1913 this duty was removed and American publishers and Canadian companies created nearly thirty major pulp and paper

Chew's mill *at Midland engulfed in flames a frequent occurrence at Georgian Bay mills c.1920.*

operations in Ontario. Most of which relied on Crown licenses for their wood supply to provide newsprint for the North American newspaper industry. By 1919 there was a well-developed industry cutting the spruce forests of the north in ways that were nearly identical to the ways in which the pine forests of Georgian Bay had been logged in the 19th century.

An Emerging Responsibility

The legacy left by the forestry industry is profound. It is witnessed by the many lessons that have been learned about the environmental impact of the large-scale clearing of forestlands and how to create a better balance between the myriad demands on the forest, including the management of Crown lands, industry needs for saw logs and lumber, pulp and paper, and the care of wildlife, water and recreational resources.

Following the Second World War public and political concerns about forestry grew. In 1946 Premier of Ontario George Drew appointed a committee to carry out a Royal Commission on forestry in Ontario. In 1952 a Federal-Provincial Forestry Agreement was initiated and completed in 1958. This laid the groundwork for forest inventories and the identification of all forest resources in Ontario. Management agreements for private, public and forest company interests were written. In 1964 another Federal-Provincial agreement established a joint research program related to a multitude of forest related concerns and issues including forest diseases and regeneration on private and Crown lands to name only two.

In 1992 forest policy frameworks for Ontario and a national forest strategy improved policies for forest production on Crown lands, forest based tourism, improved public participation, and forest investment and revenues. Out of this process came a commitment for funding for forest management and regeneration activities on Ontario's public Crown lands. Subsequent to this the Forest Renewal Trust Fund, the Forest Futures Trust Fund, and the Lands For Life Program all created a new comprehensive land use process that now strikes a better balance between protection and use of Ontario's forest resources. This process resulted in the development of more provincial parks and protected areas, a forest management and resource based policy, and finally a forest accord that guides provincial forest policies.

Throughout the Georgian Bay basin forest products continue to play an important role in the lives of its people and the industries located there. It has taken almost 200 years for government and industry to develop polices, protective measures and governance strategies for a sustainable forest industry in Ontario. This collaboration will ultimately result in the growth of forests in the Georgian Bay basin that will grow to be older than those harvested by the First Nations and Europeans of the 16th and 17th centuries.

View from above the Castle Glen Estate, looking south along the face of the Niagara Escarpment. The prominent bluff at the right is occupied by the Osler's Bluff Ski Centre.

CHAPTER

11

COLLINGWOOD AND THE BLUE MOUNTAINS: PORTRAIT OF A CHANGING GEORGIAN BAY COMMUNITY

ANDREW D. MIALL

Introduction

As related elsewhere in this book, the southern and western shores of Georgian Bay, from Waubaushene to Owen Sound and then northwestward to Tobermory at the tip of the Bruce Peninsula, are underlain by Paleozoic sedimentary rocks which have generated a distinctive landscape quite different from that of "cottage country" of the Canadian Shield and Thirty Thousand Islands areas. Not unnaturally, the recent human history of these areas has followed very different paths and the history of Collingwood is typical of many other communities along southern Georgian Bay. Agriculture began in the 1840s with settlers from Europe and industry, principally shipbuilding, developed after 1855 with the building of a railroad from Toronto. Lately, industry has all but disappeared and the area has in many ways returned to its rural roots, enlivened by an influx of retirees and seasonal migrants drawn by skiing on the Niagara Escarpment in winter and watersports in summer. This basic pattern has been followed since the mid-nineteenth century by most of the communities around Georgian Bay reflecting a move away from short-lived extractive and manufacturing industries to a more durable and sustainable economy based on Georgian Bay as a unique place.

Landscape and geology

The Paleozoic geology of the Collingwood area consists essentially of alternations of relatively resistant limestone or dolostone layers that underlie elevated bench areas across the landscape and which also form the "caprock" of the Niagara Escarpment, and much softer shales that underlie low-lying areas.

All of these rocks were formed in a shallow sea bordering the Canadian Shield to the north and form the "cover rocks" mentioned in Chapter 1. The oldest unit exposed in the Collingwood area, about 450 million years old and belonging to the late Ordovician period, is the Georgian Bay Shale which is well exposed along the shore of Georgian Bay in Craigleith Provincial Park. The shale is famous among fossil collectors for the thousands of trilobites it contains, visible to casual inspection of the outcrops.

At Craigleith, a shale-oil works was established in 1859, but the plant was not economic and was shut down in 1863. The overlying Queenston Shale (named after the village of Queenston in the Niagara Peninsula) is not well exposed in the Collingwood area. This

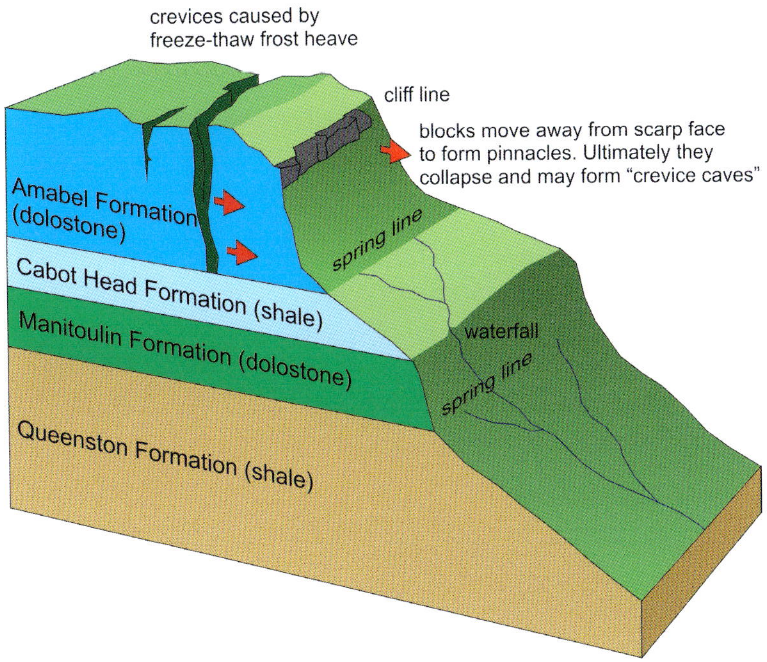

Left: ***View eastward from the top of the Blue Mountain*** *above the Intrawest Village. The now disused grain terminal at Collingwood Harbour is the most prominent feature in the near distance. In the far distance the white sand strand of Wasaga Beach is visible. The hills forming the skyline at the horizon constitute the rolling countryside underlain by Paleozoic strata in the area around Elmvale.*

Top: ***Simplified cross-section*** *through the Niagara Escarpment.*

Top right: ***Typical lake-shore exposure*** *of the Georgian Bay Shale, near Craigleith Provincial Park. Note the prominent fractures tha similarly affect rocks of the Niagara Escarpment.*

Right bottom: ***Close-up view of the numerous trilobite fossils*** *in the Georgian Bay Shale and seen along the coast at Craigleith.*

Top left: ***The Cabot Head Formation,*** *a soft red-coloured shale that is very susceptible to gully erosion. This view is at the Alpine Ski Club, just south of Craigleith.*

Top right: ***Metcalfe Rock*** *from the east, in the early Spring.*

unit and the Cabot Head Formation that occurs higher in the section are soft red-coloured shales that readily erode into muddy slopes and gullies along the face of the escarpment.

There is a gap (called an "unconformity") representing a few million years of unrecorded time between the Queenston Shale and the overlying Silurian rocks of the Manitoulin Formation, but it is not well exposed in the Collingwood area. The Silurian rocks, about 440 million years old, comprise the Manitoulin, Cabot Head, and Amabel formations. The latter is the most important, as it is the hard caprock forming the top of the Niagara Escarpment and is quarried for construction stone notably at Wiarton. The main lithology is a pale grey to white dolostone, typically containing numerous small cavities ("vugs") where fossils have been dissolved out. It consisted originally largely of the remains of small fossil "bioherms" composed of corals, and sponges called stromatoporoids, which formed reefs on the then sea floor. In some locations along the escarpment, as at Metcalfe Rock and Old Baldy in the Beaver Valley south of Thornbury, the cliff face is tens of metres high, and is a favourite of rock climbers. The Amabel Formation also forms the spectacular coastal cliffs along the Bruce Peninsula and on Manitoulin Island where escarpments have been extensively modified by glacial erosion with large re-entrant valleys cut into the face of the escarpment.

Evolution of the Modern Landscape

The rocks underlying the landscape in the Collingwood area are all more than 400 million years old when the area was covered by shallow seas. During the mid- to late part of the Paleozoic era, the sea withdrew and for most of the time since, the land has been slowly undergoing erosion. During this very long period, the Niagara Escarpment developed as consequence of the presence of the hard cap rock of the Amabel Formation which extends from Queenston in the Niagara Peninsula, to Tobermory, at the tip of the Bruce Peninsula.

In the case of the Blue Mountain near Collingwood, the Amabel Formation cap rock is underlain by soft shales of the Cabot Head formation. These shales form a kind of lubrication for large masses of Amabel dolomite forced away from the cliff face; a process geologists call "cambering." The result is the formation of wide crevices in the cap rock, and the development of blocks and pinnacles isolated from the main cliff face. Collapse of these blocks, or the filling of crevices with fallen rocks, may leave cavities below, which are termed "crevice caves." The caves in the Duncan area, south of Thornbury, and the Scenic Caves west of Collingwood, are all of this type.

At Collingwood Scenic Caves, crevices have been widened by the rotation of large blocks away from the main cliff face, leaving isolated rock pinnacles like the one called Ekarenniondi, The Watcher, or Oscotarach, the Head-Piercer. It served as a lookout for the Petun tribe, who lived there in the seventeenth century.

The interbedding of porous rocks such as the Amabel and Manitoulin formations with the less porous Cabot Head and Queenston shales gives rise to a distinctive drainage pattern along the face of the Niagara Escarpment. Groundwater seeping into the porous rocks from rainwater percolates downward under gravity until it meets the impermeable shale. It can then only move laterally, along the top surface of the shale, emerging eventually at the scarp face where the contact between the shale and the overlying limestone or dolomite crops out. Along this outcrop there may be a permanent line of springs. Larger streams that have head waters well back from the scarp face typically descend over the cliff line as a water fall. There are many such water falls along the escarpment in the Collingwood area.

Cliff formed of the Amabel Formation at Lion's Head on the Bruce Peninsula.

Limestone caves filled with stalagtites and stalagmites are not present in the Niagara escarpment, probably because the area is humid and the water table has always been too high. This type of cave requires a dry ("vadose") zone below the surface, through which groundwater percolates, dissolving limestone and redepositing calcium carbonate.

Glaciation and Deglaciation

Commencing several million years ago, the climate of Canada began to be dominated by alternations between ice ages and interglacials. Permanent ice caps formed across the continent and buried Georgian Bay on many occasions, leaving thick layers of what geologists call "till," a coarse, unconsolidated deposit consisting of an unsorted mass

Left: **The beginnings of a major crevice,** *forming near the face of the Niagara Escarpment as a result of freeze-thaw. The crevice is already filled with forest debris.*

Right: **At Metcalfe Rock, near Thornbury, a wide crevice** *has opened up from the main cliff face, and is beginning to be filled with fallen rock debris.*

of mud, silt, sand, pebbles, and boulders. Southern Ontario is covered with these deposits and now form the thick soils that underpin cultivation and farming around southern Georgian Bay. Such deposits also unfortunately include many millions of large boulders called "glacial erratics" so named for their "out of context" character. These are boulders consisting of granites and metamorphic rocks originating on the Canadian Shield country completely different from the surrounding Paleozoic sedimentary rocks.

Removing these rocks from the floor of the primeval forest, together with the removal of huge tree stumps, were arduous tasks for European settlers when they cleared the land for farming in the first half of the nineteenth century.

The retreat of the last ice sheet from the Georgian Bay area about 11,000 years ago allowed the waters of Lake Michigan and Lake Huron to merge in a large water body that geologists call glacial Lake Algonquin. Wave erosion developed prominent wave-cut platforms, beaches, and cliff lines all around the shores of Lake Algonquin, at elevations up to tens of metres above the present level

of Lake Huron. As long ago as the late nineteenth century geologists realized that these beaches do not define a horizontal surface, as do modern beaches, but that their elevations rise gently to the north. This is because of a process called "glacio-isostatic rebound" as explained in Chapter 4. The great weight of the ice sheet, nearly 2 kilometres thick over Georgian Bay depressed the earth's crust, and since the ice melted the crust has been "rebounding" much like memory foam, causing an upward tilt directed toward the area where the ice cap was the thickest, around Hudson Bay. At Collingwood the ancient

Left: ***The widening of a crevice*** *and the formation of isolated pinnacles by the rotation of blocks away from the main cliff face, lubricated by movement across the underlying softer Cabot Head Shale; Scenic Caves, Collingwood.*

Right: ***The pinnacle called Ekarenniondi,*** *The Watcher, or Oscotarach, the Head-Piercers, Scenic caves, Collingwood.*

Eugenia Falls *at the head of the Beaver Valley south of Thornbury.*

Algonquin beach is now at about 209 metres above sea level, some 32 metres above the present level of Lake Huron.

Recent and Current Geological Processes

Wasaga Beach lies east of Collingwood and is famous for its 14-kilometre-long sand beach (the longest fresh-water sand beach in the world), a favourite destination for families escaping the city heat in summer. The beach was formed by currents in Georgian Bay driven by the prevailing northwest wind, which have resulted in winnowing of glacial debris on the lake floor and from eroding coastal cliffs, resulting in the accumulation of vast amounts of sand along the southeastern shore of the lake. Some of the sand has been carried by the wind onshore to form sand dunes.

Modification of the landscape by present-day processes is slow, and occurs mainly during very rare violent weather events such as that of March 2012 at the Alpine Ski Club, Craigleith, when a drenching rainstorm occurred during a warm period just as the snow was beginning to melt. The soft mudstones underlying the base of the escarpment were saturated, and surface runoff rapidly eroded the thawed-out surface layer leading to a rapid collapse of the bank of the gulley draining down the side of one of the ski slopes. Trees and bushes collapsed into the gulley, and the mix of broken branches, boulders, mud, and water formed a debris flow, a dangerous wet mass that began moving down the slope, toward an area of ski chalets beyond. Another debris flow on the Niagara Escarpment may not happen for another century or more. But this is how the process of erosion proceeds.

Another important issue as related elsewhere in this volume is changing water levels in Lake Huron which affect docks, harbours, and marine navigation. In recent years, lake level reached a high at 177.5 metres above sea level in 1987 and fierce storms during the fall of 1986 and 1987 did substantial damage to shoreline properties, and considerable capital was expended installing sea walls and other defences. Since then water levels have fluctuated downwards reaching a low of 175.6 metres in 2013, a level reached twice before, in 1926 and 1965 since record-keeping began in 1918. Two successive cold winters, in 2013 and 2014, meant that the Great Lakes remained largely frozen over throughout the winter, resulting in less evaporation, and as a result, a strong recovery has taken place. These changes are natural and have nothing to do with anthropogenic climate change and emphasize the need for coastal infrastructure to be sensitive to changing lake levels.

Industry Arrives on Georgian Bay

Humans first arrived in the Collingwood area about 11,500 years ago just when the last ice sheet was retreating north and the waters of glacial Lake Algonquin lapped against the base of the Niagara Escarpment. As related in Chapter 7 the first industry in the area consisted of mining chert (a hard flint-like rock used to make scrapers and blades) from the slopes of the escarpment that expose the so-called Fossil Hill Formation. Collingwood was founded by Europeans in 1840 on land that had been ceded to the British by Ojibwa peoples but real development only began after 1855 when the Ontario, Simcoe & Huron Union Railway from Toronto reached Collingwood. Steamer traffic expanded on Georgian Bay and several major regional centres were established, including Midland, Penetanguishene, Port

McNicoll, Collingwood, and Owen Sound. Railways reached Midland in 1872, Owen Sound in 1873 and Wiarton in 1882.

First named Hen and Chickens Harbour by early European settlers, the town was renamed as Collingwood in 1854. With the arrival of the railway from Toronto the town became an important transit point on the route to the North American interior. Travellers and immigrants would arrive by train and transfer to steam ships providing regular service to such locations as Detroit, Chicago, Fort William, and Duluth. The first mail service to Lake Superior used Collingwood as the transhipment point in 1858. Eventually the construction of direct railroads to these locations diminished the importance of this route, but by then shipbuilding had become an important industry. A dry-dock was completed in 1882. The first terminal was built to facilitate grain shipment and the first grain elevators were built in 1871 to store

A lonely glacial erratic *in the woods near the Oslers Bluff Ski Centre.*

View of the Niagara Escarpment and the Georgian Peaks Ski Club showing the former shoreline bluff of glacial Lake Algonquin (arrowed).

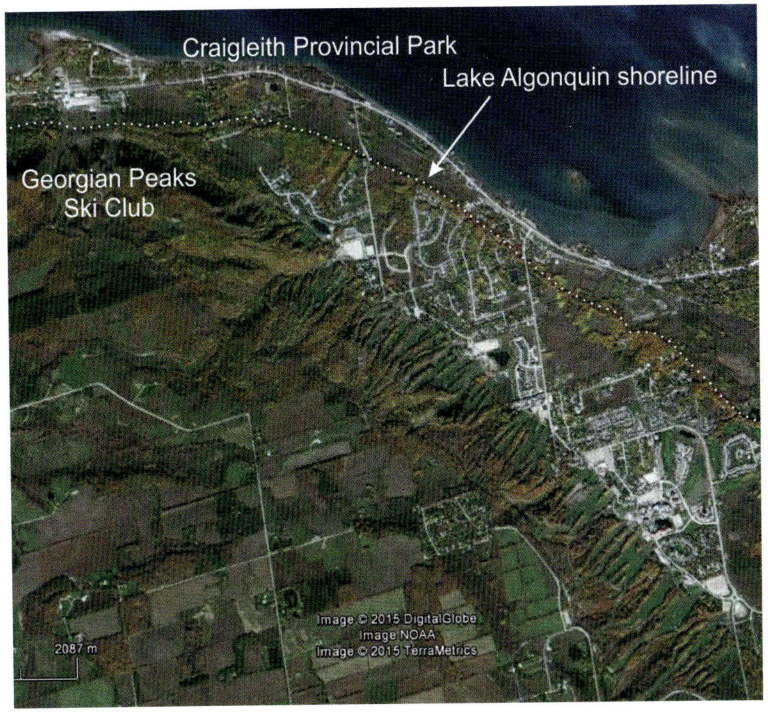

mid-west grains brought in by boat. The opening of Depot Harbour in 1897 near Parry Sound and the final construction of the Ottawa, Arnprior and Parry Sound Railway linked Georgian Bay with Ottawa and the east coast. By then, water depth limitations in Collingwood Harbour had displaced much traffic to Owen Sound which opened up regular steam traffic to Port Arthur (Thunder Bay).

The ship building industry flourished in Collingwood through the first half of the twentieth century, with the construction of some of the corvettes and other ships used by the Royal Canadian Navy during the Second World War, and some of the bulk carriers built for use on the St. Lawrence Seaway. However, by mid-century competition from overseas eliminated any local advantages that Collingwood offered, and in 1986 the shipyard was closed down and regular service on the railway terminated. Highway 69 when competed in 1955 from the Severn River to Sudbury via Parry Sound bypassed Georgian Bay and was the beginning of the end for passenger traffic on Georgian Bay. The final straw was the opening of the St. Lawrence Seaway in 1959 which led to a loss of transhipment traffic because large ocean-going vessels could now reach Lake Superior. The last scheduled boat service on Georgian Bay is provided by the *Chi-Cheemaun* which commenced between Tobermory and South Baymouth on Manitoulin Island in 1974.

Skiing came to the Blue Mountain west of Collingwood in the 1930s, and expanded steadily in importance as a local tourist industry and exploits the 220 metre relief of the Niagara Escarpment. After the closure of the shipyard the town made a painful adjustment from its role as site of heavy industry to one serving recreational needs and skiing has played a large role in this transition. The town also had to face the environmental legacy of its past. A century of ship building

Top Left: ***Aerial image*** *of the country near Craigleith, west of Collingwood. The dotted line traces the course of the high Lake Algonquin shoreline.*

Above: ***The Overhang*** *at Bruce Peninsula National Park.*

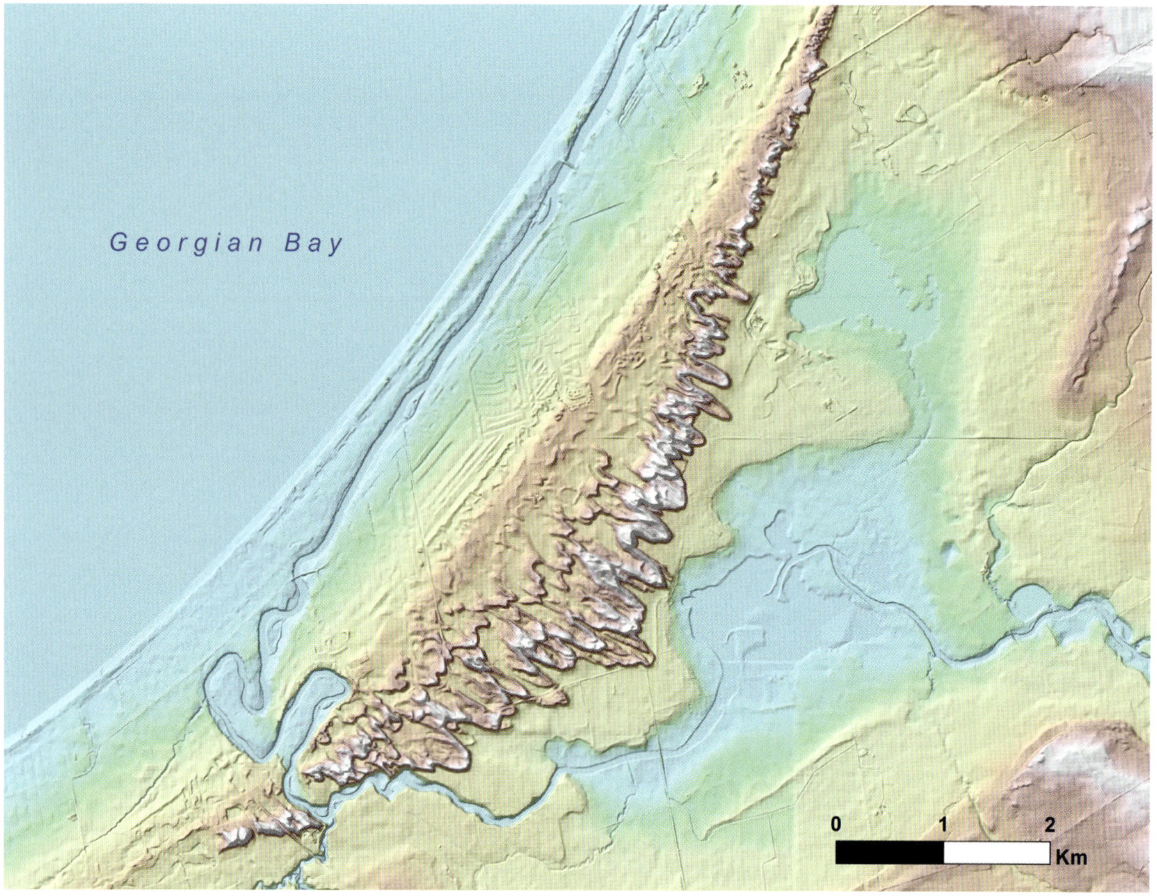

Above: ***Wasaga Beach*** *and the fossil sand dunes. The prevailing wind direction is off Lake Huron and is reflected in the evidence of downwind migration of dunes.*

Above right: ***The debris flow*** *that developed half way down the Arrowhead Trail at Alpine Ski Club, Craigleith.*

Right: ***Collingwood Harbour,*** *before and after completion of site clean-up and prior to redevelopment.*
(© HazMat Management Magazine)

activity at Collingwood harbour created a major industrial site on the shores of Georgian Bay. The closure of the shipyards and the railway in 1986 left a 16-hectare area full of industrial discards, landfill, coal ash, and soil polluted by industrial wastes, leaking fuel, and other such material. A cleanup was authorized and funded in 2004 and after buildings had been demolished and removed, the entire site was excavated, with 300,000 m^3 of soil and other materials on site undergoing sorting, testing, and partial removal. Some 4,000 m^3 of steel was recycled, plus large amounts of crushed stone and clean fill and the cleanup was completed in 2006. Today, the town has largely re-invented itself as a destination for retirees and tourists and the harbour has been redeveloped as a waterfront urban setting.

Today, access to Georgian Bay and its coast means something completely different from that prevalent 100 years ago. Formerly the Bay was simply an outlet for exporting lumber and fish and an embarkation point for immigrants moving west. The railway and the steamer ruled. Farming was the mainstay around its southern shores and manufacturing industry, as much as there was, took advantage of this basic trade pattern but left remarkably little substantive growth outside small portside communities. Changing demographics since the end of World War II, especially longer living urban populations, better road transportation and proximity to Toronto and the surrounding urban areas, together with the opportunities created by the World Wide Web, have turned the Bay into a desirable destination ushering in a new phase of urban growth around its shores. Urban development of waterfronts on a scale not seen prior, creates entirely new challenges in regard to conservation and protection of the Bay's waters and ecology.

2006
1956

Ship building at Collingwood, using the famous 'side launch' method. The shipyards built more than 200 ships including corvettes for the Royal Canadian Navy during WW II. The last ship, completed in 1986, was the Canadian Coast Guard vessel CCGS Wilfrid Laurier (All photographs courtesy of Collingwood Museum).

ENGLISH RIVER
CANADA

Launch of Str Waubic Collingwood
June 1-1909
907

F.H. Varley, Stormy Weather, Georgian Bay, 1921, oil on canvas, (National Gallery of Canada, Ottawa)

CHAPTER

12

ART OF GEORGIAN BAY

CHRISTINE BOYANOSKI

***Alan Stein**, Winter at the Open, Dusk #2 (Courtesy of the Artist).*

*Far right: **John William Beatty** (1869-1914), Fozen Lake in North Ontario (Courtesy of the Art Gallery of Ontario).Beatty was a prolific painter of landscapes across the Parry Sound District of Georgian Bay, especially in what is now Algonquin Park. He was an early influence on the Group of Seven.*

Georgian Bay has captured the imagination of artists for generations and has been immortalized in the paintings of the Group of Seven; classic works such as F.H. Varley's *Stormy Weather, Georgian Bay,* 1921, have helped shape popular perceptions of the region in the minds of many Canadians until it has come to stand for the northern wilderness and Canada itself. The wind-tossed pine trees in Varley's painting—metaphors for the hardy Canadian spirit which triumphs over adversity—became a familiar motif in the Group's repertoire.

In reality, Georgian Bay was hardly the empty wilderness depicted in their canvases. The area was populated by First Nations

237

Georgian Bay

Patterned Rock and Aura, *Wreck Island*
2016. Photograph by John Krasinski.

people which earlier artists like Anna Jameson (1794-1860) and Paul Kane (1810-1871) had recorded in the sketches they made while travelling through the region in 1836 and 1845, respectively. Later in the nineteenth century, the logging industry became a visible presence in the landscape, as did the growing number of visitors who built summer retreats at Georgian Bay, seeking "an authentic wilderness experience." The Madawaska Club was one such group. Founded in 1898 by affiliates of the University of Toronto, the Club initially aimed to combine scientific research with recreation around land it had purchased at Go Home Bay.

Photographs taken in this area by three of its members around the turn of the twentieth century reflect how cottagers saw the Bay and what they valued about it: a place of social intercourse and recreation and an environment in which they could appreciate the beauties and curiosities of nature. This was the Georgian Bay to which J.E.H. MacDonald (1873-1932), Arthur Lismer (1885-1969), A.Y. Jackson (1882-1974), and Tom Thomson (1877-1917) were first introduced between 1912 and 1914 when they visited as guests of Dr. James MacCallum, a Madawaska member and patron of the arts. Thomson preferred the solitude of Algonquin Park to the social whirl he encountered at Go Home Bay. Nonetheless, Georgian Bay was one of the early locations that provided these artists with an appropriate landscape through which to realize their vision of Canadian art. It was a northern "wilderness" landscape, far removed from the artificiality of the city, onto which their notions of cultural nationalism could be projected. Appropriate views were found that were either free of human traces or were edited to erase them and then manipulated as required. Lismer painted the sketch for *A September Gale, Georgian Bay*, 1921 near the spot where Varley was painting his *Stormy Weather*, but his treatment yielded a different result.

Group members like A.Y. Jackson continued to visit Georgian Bay during the decades that followed their heyday in the 1920s. Lismer, whose teaching schedule restricted his sketching trips to summer vacations, would return repeatedly to McGregor Bay and Manitou Dock until the early 1950s. His new emphasis on the formal properties of his paintings (colour, line, form) and close-up focus on the lichen-covered rocks, roots and other foreground details beginning in the 1930s marked a change from his earlier work at the Bay.

Doris McCarthy, *Georgian Bay, 1968* *(Courtesy Doris McCarthy Gallery, Scarborough College, University of Toronto).*

After the Group of Seven

The Georgian Bay paintings of Tom Thomson and the Group of Seven provided precedents for artists who followed their example by venturing into the region to paint the landscape. In many ways, the Group taught people how to see the Bay. Charles Comfort (1900-1994) acknowledged the importance of their legacy in a 1951 article written for *Canadian Art* magazine. He also explained why artists were attracted to the area, writing:

"About these harsh barren islands there is an undeniable fascination, possibly inherent in the manifest struggle recorded in their

Alan Stein, *Night Swimmer at Bayfield Inlet*. Wood engraving, 1998, an original illustration for the poem of the same name by Douglas LePan, from the limited book, *Towards The Open*. (Courtesy of the Artist)

Right: **Ed Bartram**, *Island Forms* 1974. Ed Bartram, *Precambrian Arc*, 1986 (both Courtesy of the Artist).

every line, perhaps in the sense of remoteness and stern resistance to the encroachment of man...The crystal-clear waters of Huron lap the shores, revealing countless rocky abstractions beneath."

Some of Comfort's sketches made at Georgian Bay in the late 1940s and early 1950s are clearly indebted to the Group. The large canvas, *Split Rock, High Noon*, 1950, however, is characteristic of Comfort's more abstracted rendering of nature. The simplification of the rock face into bands of varied colours is consistent with his written description of these forms: "low whale-backs of fundamental strata, describing in their convulsed fractures the heaving turmoil of creation...." He may have been thinking of the wartime images by British war artist Graham Sutherland and others which recorded the twisted girders of buildings demolished by wartime bombing—a heaving turmoil of destruction.

Doris McCarthy (1910-2010), who had been a student of Arthur Lismer and J.E.H. MacDonald at the Ontario College of Art, established herself in the Toronto arts community soon after graduating in 1930, becoming a full-time teacher at Central Technical School in 1933 and a member of the Ontario Society of Artists in 1945. McCarthy, a landscape artist, had visited Georgian Bay in 1935 and 1939, but it wasn't until 1959 that she began making regular visits—renting, then buying a cottage-studio at Key Hole Bay which she described as "an escape to peace" in 1969. While her early work was strongly influenced by the Group of Seven, McCarthy began to experiment with different forms of abstraction that were current in the 1960s. In *Georgian Bay* of 1968, the characteristic elements of the landscape are present, but in "hard edge" works like these, they have been analyzed, then reduced to a set of rhythms rendered as flat areas of colour. These paintings appear to have been inspired by the view made possible by flying over the region which would have revealed the large patterns of the rocks and water below, aiding the process of abstraction. In her later work, McCarthy returned to a more representational treatment of the landscape.

Goodridge Roberts (1910-1974) brought a different approach to painting the Georgian Bay landscape. An established Montreal artist, Roberts was known for his landscapes of the Laurentians and Eastern Townships of Quebec, as well as for his still life and figure paintings. He first visited Jackknife Lodge, located north-west of Parry Sound, in 1951 when Georgian Bay was still strongly associated with the Group of Seven. Vince Thomas, a former student, had invited him to the Lodge which he managed and which was frequented by artists. Apparently, Roberts initially observed that all the trees at Georgian Bay had been planted by the Group of Seven—presumably those wind-tossed trees clinging to bare rock—but this is not how he painted them. Having absorbed the basic tenets of French modernism from his teachers at the Art Students League in New York, Roberts'

15/50
Island Forms
Northern Image Series
Bartram /74

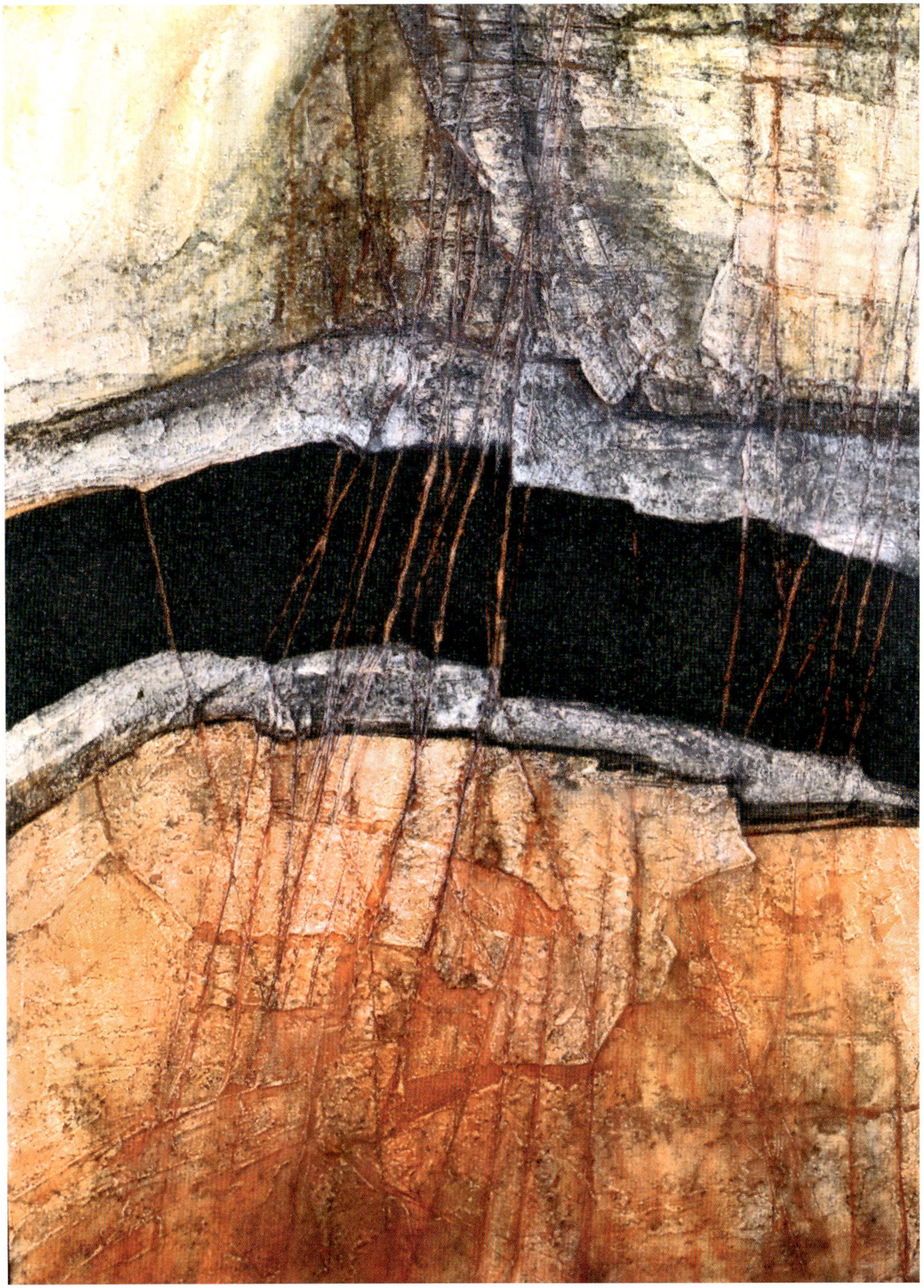

Black igneous intrusive dikes *cutting through lighter banded gneiss and overlapping crystal clear blue waters are reflected in placemats by Parry Sound quilter Diana Thompson.*

Right: ***Monica Tap,*** *Wake near Honey Harbour, 2010 . (Photo by Gordon Hicks; courtesy of the Artist).*

aim was to create an aesthetic experience in which each part of the picture was related to the whole. In *Georgian Bay* of 1955, painted directly from nature, this unity has been achieved through the texture of freely applied paint and the harmonious balance of rich colours which also characterized his Quebec canvases of the same period. The result is a landscape suffused with light and air. Roberts would return to the region every year thereafter, with the result that he became identified as a painter of Georgian Bay.

A unique perspective of Georgian Bay was provided by Angus Trudeau (1908-1984). An Odawa from Manitoulin Island, he worked for years as a cook on commercial vessels in Lake Huron. The paintings and models of the boats that he made in his spare time and after his retirement record his personal experience of life in the region. In mixed-media paintings like *Manitou, Tobermory, Ont*, Trudeau, a self-taught painter, often incorporated clippings from contemporary newspapers and magazines which he glued directly onto the surface of the work. Exemplifying the fresh approach of a "folk artist," Trudeau was picked up by the Isaacs Gallery of Toronto in 1973. According to gallery owner Av Isaacs, Trudeau was a chronicler of life on Manitoulin Island; and he was an important influence on younger First Nations artists on the Wikwemikong Reserve where he was an elder later in life.

Contemporaries

The legacy of earlier artists is undoubtedly a reference point for many contemporary artists who work there. These artists know the region well and most have a long-term attachment to it. Some live there year-round or return every summer to cottages, many of which are accessible only by boat. It can be an inhospitable environment with its exposure to storms and shoals but the "Bay people" are committed to what they view as a more authentic wilderness experience, distinct from life in more protected cottage country like Muskoka.

Anong Migwans Beam (b. 1980) is an Ojibwe artist who lives and works on Manitoulin Island. From her studio at Kagewong overlooking Mudge Bay, she produces large paintings in oil and photo transfer, and monoprints which express her personal philosophy through the powerful image of water. This work bears the influence of her life on the island with its strong Native presence and natural surroundings and of her parents, artists Carl and Ann Beam. Anong believes that the past and future are connected in the present moment of human existence, writing, "the structure of time does not limit us, and all moments can occur simultaneously. So that everything, past, future, is currently unfolding in a flowing way in the present moment. The static image of flowing water is how I have come to portray this understanding."

Ivan Wheale (b. 1934), a self-taught painter, settled on Manitoulin Island in 1975. He finds subjects for his high realist canvases in areas of limited access in the North Channel of Manitoulin from Blind River to French River. Working from photographs, Wheale invokes a sense of drama through his manipulation of lighting and viewpoint. Typically, a rock face viewed at close range confronts the viewer at a

***Angus Trudeau**, (1908-1984), mixed media on paperboard (Courtesy of McMichael Canadian Art Collection).*

dramatic angle, lit in such a way as to focus the viewer's attention on the colour and texture of the granite or limestone. Wheale uses the standard components of a Georgian Bay landscape to construct works in which the rocks, trees and light realize their potential to make a strong impression on the viewer.

Bruce Steinhoff (b. 1957) has been visiting Georgian Bay for well over twenty years, paddling and tenting up and down its eastern shores from Port Severn to Britt, making small oil on birch sketches in the tradition of the Group of Seven. For Steinhoff, it is imperative that he have a connection to his subject which he believes can only be achieved either by living there or making regular visits. He is a "naturalist" painter, recording moments of visual perception in front of his subject, rather than analyzing the view or shaping it to conform to some preconceived idea. The loosely-brushed canvases appear uncontrived, and retain their immediacy when transferred to a larger format. Steinhoff's affinity for his subject allows for the authentic translation of his perceptions through pigment on canvas.

A personal history of Georgian Bay underpins the work of John Hartman (b. 1950). Born in Midland, having resided in Tiny Township since 1981, and with an island cottage south of Britt in the eastern archipelago, Hartman considers Georgian Bay to be his home landscape. According to the artist, Georgian Bay has defined his sense of space, light and scale. In the 1990s, he began inserting narrative elements into aerial views of the landscape, filling both lower and upper registers of the picture with figures acting out events (sometimes cryptically). The landscape became a mnemonic field onto which people and events from both the collective memory and the artist's personal history and memory were inscribed. At that time, he relied on pencil sketches for the composition, as well as his memory of the light, colour and general experience of the location. However, since the summer of 2014, Hartman has been using a drone to take aerial photographs of the area from a distance of about three hundred feet, whereas previously, he had to imagine how a particular shoreline would look. He is now able to combine the vast amount of information contained in the digital photograph (whose composition he is able to control as he shoots it) with his knowledge of the area, acquired by over forty years of looking and painting at ground level.

Right: Alan Stein: Trestle Bridge, Parry Sound (Courtesy of the Artist).

Bianca Perren, *View from E19, May 2008 (Courtesy of the Artist).*

Says Hartman, "it is a thrill for me to see a place that I already know intimately but in a new way."

Ed Bartram (b. 1938) has commented that his entire art career is associated with his rocky island on Georgian Bay which he bought in the late 1960s. His work as printmaker, painter and photographer has been influenced by the banded metamorphic gneisses found in the area of Manitou Dock where his island is located. In order to avoid repeating the compositional structures used by the Group of Seven, Bartram decided to focus on the rock surfaces with their complex colours and textures, first through the medium of colour etchings. He thinks of the process of etching—involving scraping and biting of the metal plate—as a reenactment of the violent forces that first created the rocks. *Of Island Forms* of 1975, Bartram has written:

"The print...is from the Northern Image Series, a group of photoetchings of imagery from the archipelago. To make this print, I have etched on a metal plate, a collage of imagery including: an aerial map of Moose Deer Point with its surrounding islands to create the sky; and a negative image of the rock surface with its floating lichens.

Robert Murray, Kawaatebiishing, 2003, cedar, aluminum, steel (Courtesy Barbara Edwards Contemporary and thh Artist).

Above: **Joseph Hartman**, Car and Overturned boat, Norgate Inlet 2012 (Courtesy the Artist and Stephen Bulger Gallery).

Right: **Georgian Bay Pine** 2016. Artwork created by Lawrence Onisto and Cameron McCuaig of TreeCut Art from a White Pine log at least 500 years old. It was recovered by Midland-based AquaTimber, a company led by Mark and Peter Hamelin, who manufacture environmentally-friendly wood products from ancient old growth timber salvaged from the floor of Georgian Bay.

The islands and clouds are reflected in the water and everything in the image celebrates the resonance of recurring forms in nature...."

It was through the etchings that Bartram worked out his approach to painting the Georgian Bay landscape. The paintings followed later in the 1980s. The colours and shapes in his "paintings" are borrowed from nature; upon a textured ground of acrylic medium and sand, the colours are poured and dripped, and black amphibolitic sands gathered from the beaches are incorporated, replicating the sparkling effect of the sun on the black striped rocks such as in *Precambrian Arc* of 1986. These works hover between realism and abstraction.

The works of art discussed so far clearly demonstrate the many different ways of seeing the Bay and of giving that experience material form. For Alan Stein (b. 1951) Georgian Bay is the "landscape of his soul." In the late 1980s Stein and his family settled in Parry Sound where they established an independent bookstore and Alan began printing limited edition art books on his printing press. During the summer, they spend time at a cottage in Bayfield Inlet, north of Pointe au Baril. Describing himself as a painter and printmaker, Stein works primarily in pastels and charcoal and practices wood engraving and lithography. He exaggerates form and colour in his work in order to express the spiritual and mystical qualities that he finds at Georgian Bay. This strong design element gives the work its distinctive quality; he admits that it is a challenge to make the Georgian Bay work "look like your own." A winter scene, such as *Winter at the Open, Dusk #2*, 2011 is particularly effective. One of the most imaginative examples of Stein's approach to the Georgian Bay landscape is the set of wood engravings he made to illustrate Douglas LePan's poems about Georgian Bay, exemplified by *Night Swimmer.*

For watercolourist Ann MacIntosh Duff (b. 1925), Georgian Bay has been a part of her life for as long as she can remember. She explains that her parents spent their honeymoon there, and that she learned to walk on the dock at Skerryvore. It is a source of inspiration and a place of solitude where this independent spirit can pursue her art undisturbed at the island cottage her parents bought in 1944. Duff spends most of every summer there, near Pointe au Baril, as she has for years, and describes what it means to her: "My small Georgian Bay coastal island is a microcosm of nature which is moving and changing imperceptibly and constantly. From violent sweeping storms to still, mysterious, starry nights, I see something that I love so much and from which I attempt to catch the essence with paint and brush on paper."

For Tim Zuck (b. 1947) making art is a meditative process, one that requires time, quiet and thought—not surprising, considering his background as a conceptual artist. Being on the island that he bought over twenty years ago near Norgate Inlet takes him "out of the world." He believes that this isolation releases him to see things differently and be receptive to things that he might otherwise miss. The viewer's full comprehension of the painting similarly requires thoughtful and prolonged looking to grasp the formal relationships that the artist has set up. For example, the navigational marker, which is the focus of *Beacon*, 1993, attracted Zuck with its strength and its quality as a "timeless icon of stability, a straight thing in the midst

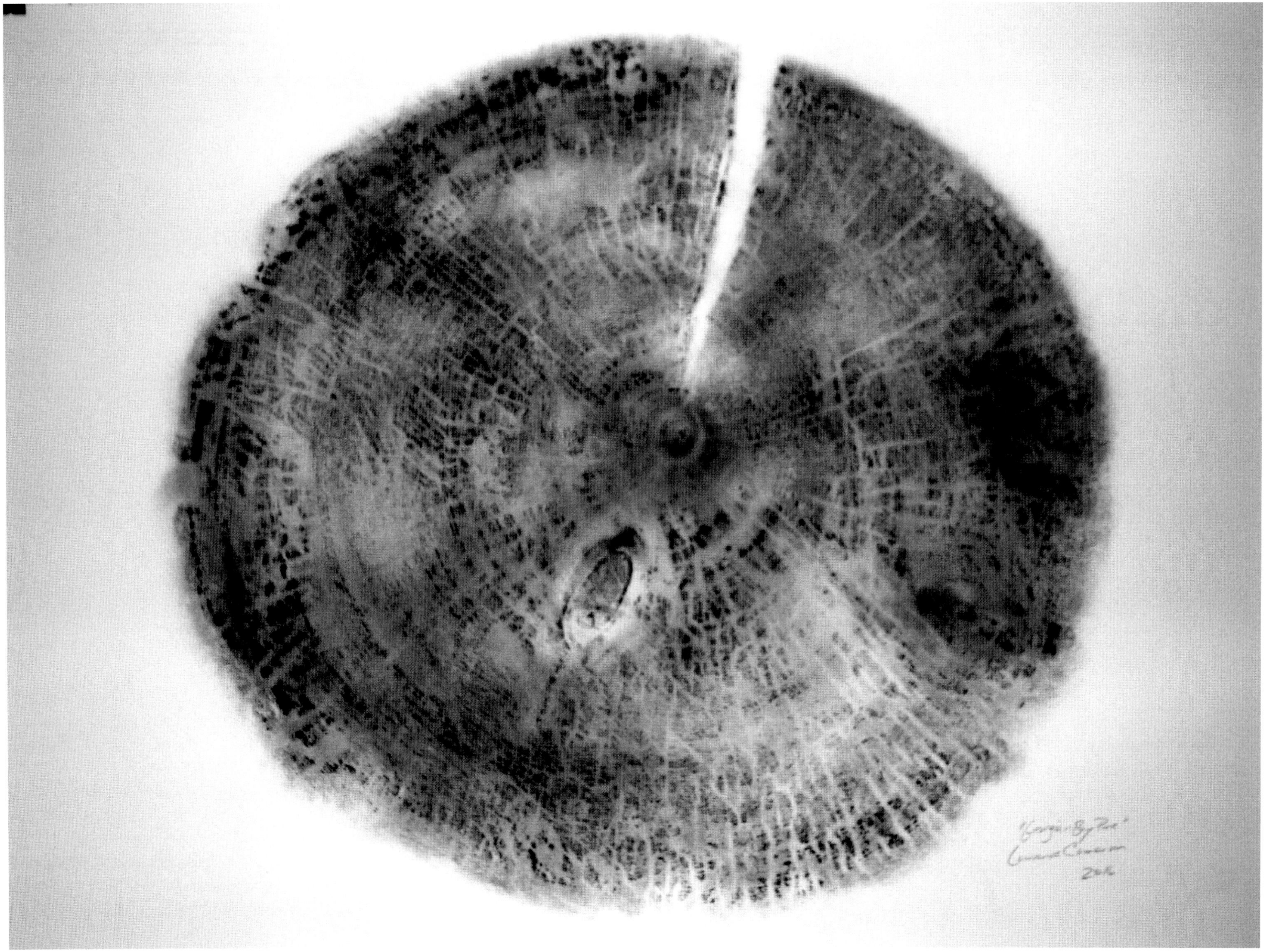

This quilt is a mosiac *of numerous small pieces of fabric just like the crustal pieces brought together by plate tectonics a billion years ago. Its pattern mimics the high mountain belts formed by the collision and it lies on now-exposed gneissic rocks of those mountains seen here on the Pancake Islands. Courtesy Diana Thompson.*

of all this potential chaos." It is not so much the fact of the beacon that Zuck wants the viewer to grasp, but the felt sense of strength it imparts in the viewer. All the formal elements work to that end: the centering of the object on the vertical axis; its elevated position; its visual stability increased by the afternoon shadow; and the weight given to the lower portion of the picture through the use of darker tones. The beacon's rectilinearity stands in contrast to those classic paintings of Georgian Bay in which windswept pines, bent by the west wind, are silhouetted against the northern sky. No storms, no drama disturb the all-pervasive calm that emanates from this work.

Sculptor, painter and printmaker Robert Murray (b. 1936) grew up in Saskatoon and recalls that his earliest notions of what constituted art were formed by paintings of the Group of Seven which he admired ("tortured pines, jagged islands, rocks...."). One of Canada's premier sculptors, Murray now lives in Pennsylvania and spends the summer months at his cabin on Lookout Island near Pointe au Baril. It is in this environment—about which he is passionate—that he conceptualizes his sculpture. Some pieces have been built for and installed on the islands of other Bay people. The work relates to the environment of Georgian Bay in a number of ways. For example, Murray makes watercolours of the Georgian Bay landscape on painting excursions in his seaplane."This work is indirectly reflected in my sculpture," he observes,"not in any literal sense, but as forms and forces in nature that become the themes of my abstract sculpture. Sometimes, the work relates to the history of the place as in *Kawaatebiishing* whose name and cylindrical forms refer to the first beacon (a modified barrel) at Pointe au Baril. Murray loves to see the contrast between his brightly coloured enamel and steel forms and the organic-looking forms of the rocks—rugged and exposed. The site itself is important to consider for the interaction between the viewer and the work of art.

Bianca Perren (b. 1975) grew up spending her summers in and around Pointe au Baril, and has always been fascinated by the shapes, colours and contours of the coastal islands. She is also a geologist who studies the evolution of landscapes and speculates that it was the beauty of the Georgian Bay rocks that got her started and that in some ways, it's as a geologist that she paints them. She writes:

"On the edge of the open, with little vegetation, you can clearly see the paths of former ice in the rounded whalebacks, the chatter marks, and also in the shapes of the islands, how they flow. You can also see their Precambrian history in the lovely pink feldspars and veins and dykes that traverse them.

"What you see now, and what I paint, is, I think, a kind of intersection of its modern self with its glacial and geological past. Watercolour...lends itself well to the task, as you can layer it and play with the translucency of those layers and still have this overall calm light feeling from the painting as you get from the land."

If Robert Murray sees the forms and forces of nature in the Georgian Bay landscape, and Bianca Perren sees its geological past, Charles Meanwell and Monica Tap see its abstract possibilities. Meanwell (b. 1946) turns his back to the Bay (perhaps because it is the view most often chosen by others) preferring to paint the transitional space between the water and the interior. He has been visiting McGregor Bay regularly since childhood, and is greatly affected by that landscape and the region just behind it. His use of a restricted palette and paring down of forms, reduces the landscape to its bare essentials and captures its abstract qualities, as in *Blueberry Island*.

Monica Tap (b. 1962) was seeing the Bay with fresh eyes when she was invited to paint a Georgian Bay landscape for the Georgian Bay Land Trust auction in 2010. *Wake at Honey Harbour* , 2010 evolved from an evening boat ride with her hosts to Honey Harbour. While shooting a video from the boat (she works from photographs and videos) Tap became mesmerized by the reflections of the landscape in the water, which appeared to liquefy, "glowing like molten metal in the soft waves created by the boat." Tap decided to make this the focus of the painting. Back in the studio, she worked from the individual video stills of this "landscape in motion," which in low resolution appear abstracted. The title refers to the wake of the boat in which the landscape was reflected, and to the experience of waking up to the incredible Georgian Bay landscape the following day.

Many artists, like Monica Tap, make use of technology in the process of producing work, but for others, photographs, films and videos constitute the work itself. Joseph Hartman (b. 1978) is a self-taught photographer who grew up near Lafontaine in the southern Georgian Bay region. Spending summers near Norgate Inlet and working as a coast guard while he was a student, he developed a personal connection to most of the places he has photographed in his Georgian Bay series. Edward Burtynsky, whose work Hartman admires and with whom he apprenticed for four years, advised his protégé to stay away from the pristine landscape as subject matter as it was tricky to make original photographs from nature. In contrast to the way the Group of Seven approached the landscape, Joseph (his father is John Hartman) believes that the truly"untouched landscape" no longer exists. He is interested in capturing the interaction between humans and the landscape, and how each influences the other. People in the region still make their living from fishing and tourism, but Hartman conveys the impression that the landscape will endure largely unchanged long after we are gone. *Car and Overturned Boat, Norgate Inlet,* 2012 juxtaposes the transience of human life—the two outmoded methods of human transportation—with ancient and enduring rock forms.

In conclusion, Georgian Bay provides a place where artists can record their perceptions of natural forms, experiment with new concepts and approaches, and where human creativity is given the space and time it requires to flourish. The region has a rich artistic heritage which some have followed, some have rejected and to which others have responded. Life on an island in the archipelago is not easy—"you have to love it," as one artist told me—but the rewards are worth it.

Artwork in nature – *folded and refolded gneisses such as these exposed on the eastern shores of Georgian Bay are a continuing source of artistic inspiration.*

Bruce Peninsula National Park

CHAPTER

13

CONSERVING THE UNIQUE ECOSYSTEMS OF GEORGIAN BAY

BILL LOUGHEED

Scientists and students at work *in the Biological Station on Georgian Bay, circa 1905. A large aquarium in the centre of the laboratory, which by 1903 was continually fed with fresh water, was critical for scientific studies, including determining conditions optimal for hatching.*

Wild places of beauty, pristine ecosystems, and solitude are hard to find close to major population centres. However, significant portions of Georgian Bay still remain essentially wilderness having now largely recovered from the ravages of earlier logging activity. As we've seen, Georgian Bay is significant as an ecological border zone between the Canadian Shield and the more fertile lowland margins to the south. Indeed, its ecological diversity is broader than either. Not surprisingly then, Georgian Bay is the only freshwater body in the world to have two UNESCO World Biosphere Reserves: one comprising the Bruce Peninsula and the other along the eastern shore (and inland lake systems) from Port Severn in the south to Killarney. Georgian Bay visitors, communities, and governments can (mostly) be proud of the success of conservation efforts to date. In this light, this review will briefly examine those efforts, as well as the regulations and plans already in place to assist wilderness preservation, and the major challenges for the future of this unique place.

The First 100 Years

The Scotsman John Muir (1838-1914) was one of the United States' most famous naturalists, an early proponent of the need for national parks, a founder of the Sierra Club, and a man forever linked to the preservation of Yosemite and the giant redwood (*Sequoiadendron giganteum*, the world's largest tree) in California. He also spent much time around Georgian Bay. Fleeing north from the American Civil War, he lived along the Bighead River Valley near Meaford, Ontario between 1864 and 1866. From here he explored much of the Niagara Escarpment along what is now the Bruce Trail. This was the time of greatly increased Ontario grain exports to the United States – a boon to the local economy but at the expense of local forests. The sights and smells of the "smoke of magnificent forest trees" being torn down and burnt to make way for cropland would influence Muir the rest of his life. A plaque at Epping Lookout on the edge of the Escarpment, on Grey Road 7 overlooking the Beaver Valley, celebrates Muir's time in Georgian Bay.

The desire to protect natural areas is a hallmark of any civilized society but effective ecological conservation must necessarily start with knowledge of what is worth protecting. This entails studying and inventorying all species of plants and animals present and analyzing their habitat. It involves measuring the prevalence or rarity of species. In that regard, Georgian Bay got off to an excellent start. In 1901, there were only three biology field stations in all of Canada. One was on the Pacific coast and one on the Atlantic. The third, the Georgian Bay Biological Station, was the first ecological research station anywhere on the Great Lakes.

It began in the fall of 1900 when a petition was presented to the Federal Minister of Fisheries asking for assistance for the proposed biological station, indicating that the chief study would involve the propagation of bass. Situated some 25 kilometres by water north of Honey Harbour at Go Home Bay, professors, scientists, and their students performed the very first ecological studies of this coastline's flora and fauna and also performed daily environmental monitoring. The scientists, mostly from the University of Toronto, travelled to this remote community by coal-fired steamers such as the *Waubic*. Their residency usually lasted from May 15th until the third week of September. This was a hardy breed of dedicated scientists, not intimidated by wilderness, who dragged along with them their most promising students.

Mail and supplies were collected by rowing out to the steamer as it passed by these wilderness shores or by collecting weekly supplies, often by rowboat, at the steamer dock some 3 kilometres distant. Dr.

R.R. Bensley, the first director of the station, described the laboratory as follows in 1902:

> *It has an ample floor-space of 20 x 40 feet, and is divided into a general laboratory and four small rooms, the latter being used respectively as photographic, glassware, chemical, and private rooms. It is provided with excellent equipment of chemical reagents, glassware, special apparatus and chemical balances. The main laboratory is provided with twelve work-desks, which are accommodated in and between the window spaces, the centre of the room being occupied by a large aquarium-table, with water connections, so that aquatic forms may be kept alive while under observation.*

The research station was a centre for serious academic research, funded at today's equivalent of $35,000 per annum which was indeed a good investment. For instance, as early as 1893, four hundred commercial fishermen operated on Georgian Bay with over 150 boats, 15 tugs, and 1.5 million yards of gill net. The station's research data had many important uses, not the least of which was to inform and set commercial and sport fishing regulations. By so doing, the research performed at this small remote station played a critical role in saving many Georgian Bay fish species.

Baseline studies done by the fledgling Georgian Bay Biological Station provided the first systematic study of the flora and fauna of the region where species were identified, microscopically examined, dissected, categorized, and further studied. As we know, these types of baseline studies are a critical and necessary step in species conservation and habitat stewardship. Despite the introduction of commercial fishing in 1835, circa 1905 there was an "absolute abundance of fish" according to the scientists at the station, so much so that specimens for biological study could be collected "by the simple use of hoop-nets, hand seines and cheese cloth nets (for fry)." One of the laboratory's first endeavours was observing and understanding the spawning habits of largemouth and smallmouth bass. The controlled scientific documentation and quantifying of bass nesting behaviours was responsible for moving the opening day of bass season to July 1st. According to Professor B.A. Bensley, this probably helped save this population.

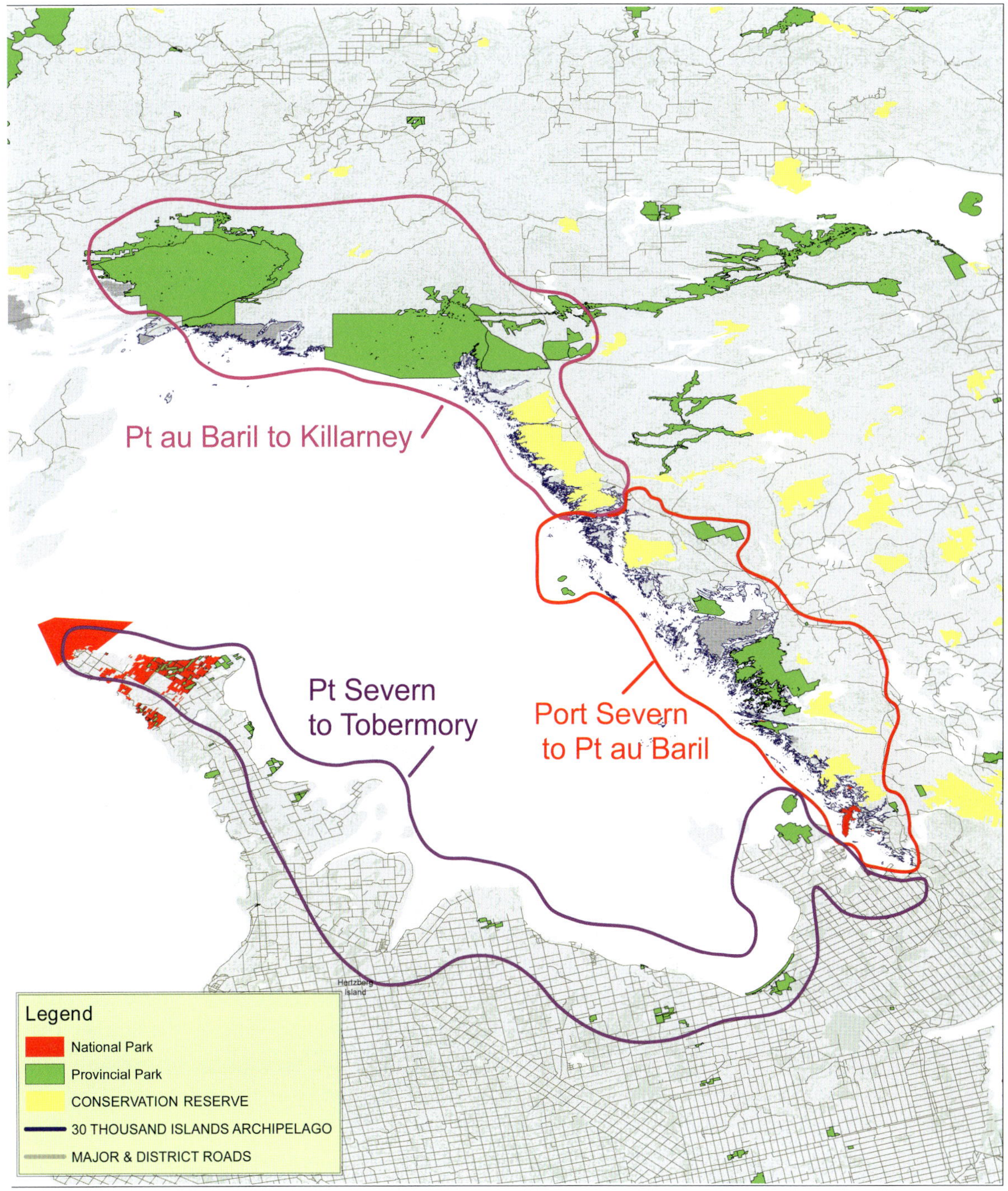

***Georgian Bay areas preserved** by Provincial and National Parks, and Conservation Reserves. Outlined areas are "Natural Areas" defined by the Nature Conservancy of Canada and the Georgian Bay Land Trust where conservation efforts are focused.*

Some parts of Georgian Bay's eco-system were protected very early on in the province's history — but only for utilitarian purposes. The eastern white pine *(Pinus strobus)* is a good example. Classically associated with Georgian Bay largely because of its imposing size, the white pine is often seen towering above surrounding hardwoods. It can also grow straight out of rock, a habit much captured in the iconic paintings of the Group of Seven and others that emphasized the unusual, often lopsided wind-blown crowns of these trees whose shorter branches point upwind. The white pine was adopted in 1984 as the provincial tree of Ontario given its importance in the history of settlement. It is fast growing (sometimes a metre a year) and can live for several hundreds of years. These trees were originally conserved for exclusive use by Britain's Royal Navy for masts and spars, and were often marked by surveyors with the King's Broad Arrow (three hatchet slashes made on the trunk). Resentment against this edict among American colonists led to the Pine Tree Riot, and together with the taxation of tea culminated in the American Revolutionary War. In Ontario, such regulation of these trees lasted until 1854. Thereafter, white pine became more valuable in the cut-lumber trade that was linked to a rapidly urbanizing Great Britain.

The ability of pine trees to grow out of rock on Georgian Bay's pine barrens – so typical on the many wind-swept and otherwise bare islands of the Bay – is due to breakage of old hard rocks by stress. All rocks, no matter how hard, are cracked. Geologists call these features fractures which form in response to complex compressional stresses imposed by the movement of the North American plate. They are also caused by extensional forces resulting from uplift (rebound) of the Earth's brittle crust following the last ice age when the great weight of the ice sheets melted. Fractures play a major role in controlling the direction in which rivers flow (notice the many straight rivers on the Shield). They also store huge quantities of snowmelt and rainwater which allow the higher standing Shield hills to act as giant rain barrels. Much of this water is fed as groundwater that flows through complex fracture systems to discharge on the floors of lakes and ponds. In a long hot summer, the water levels in the rock will be drawn down, often killing off trees and shrubs. In fall, fractures will begin to refill. Little noticed and seldom discussed, fractures and their associated waters are a vital part of Georgian Bay's ability to support a diverse ecosystem on hard rock.

In 1902, comparative studies were made of adult fish and their feeding sources (stomach contents). The young and immature of each species were examined for growth rate studies. In 1903 research was conducted on roe and identification of species-specific predators of various roe. In this regard, Bensley describes the hatching pond developed by the station: "A hatching pond for bass was last fall prepared, and twenty-nine bass of different ages were placed therein. If they winter successfully, their spawning operations are to be carefully studied. The preparation of another hatching pond for bass is contemplated next season that the effect of food and environment may be determined, as well as the growth from year to year."

Over many years, the station performed work in many areas of science. A meteorological station was established with daily measurements taken of precipitation, wind speed, wind direction, and temperature. A hydrological survey of the area was conducted. In-depth studies were conducted on aquatic insects, crustaceans, molluscs, flowering plants, as well as migratory and resident birds. Faunistic studies looked at inventories, habitat, and behaviours. Entomologist Dr. E.M. Walker performed the first systematic studies on the insects of the Georgian Bay coast and later received the Royal Society's Flavelle Medal for outstanding scientific achievement.

The work at the biological station reflected the evolution and emergence of the scientific discipline of ecology in the early 20th century. Inventories of flora and fauna conducted from 1902 through 1913 laid the basis for later studies but it was not until very recently that this job was completed by comprehensive field work and reporting by Jalava et al. (2005). Their *Ecological Survey of the Eastern Georgian Bay Coast* published by the Nature Conservancy Canada, represents a milestone in understanding of the natural ecological heritage of the area and is a foundation for future conservation activities.

On the other hand, actual land conservation got off to a slow start on the Bay. The first initiative occurred in 1929 when the 8-kilometre-long Beausoliel Island became a National Park that was later expanded and named Georgian Bay Islands National Park. From 1900 through 1950 very little else happened in setting aside wild places, but then suddenly land conservation began exploding as Georgian Bay fell within easy road access of Toronto in what has been called "the great rush to get outdoors." From a mere 1,350 hectares (ha) in 1950, conservation lands grew to 8,500 ha in 1960 and then to 59,000 ha by 1980.

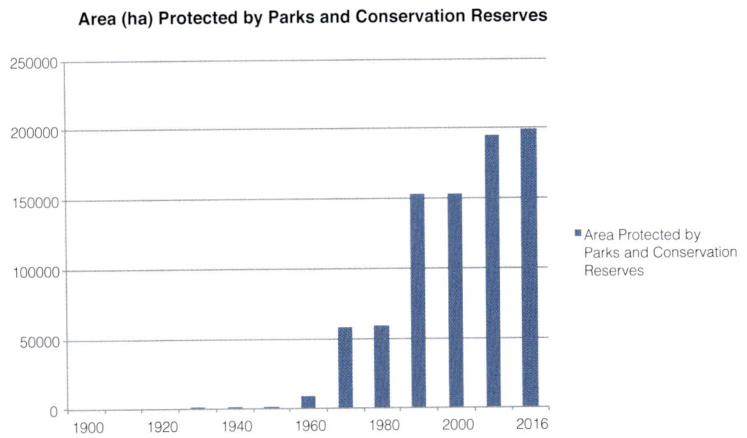

Area of conserved land over the last 100 years in Georgian Bay

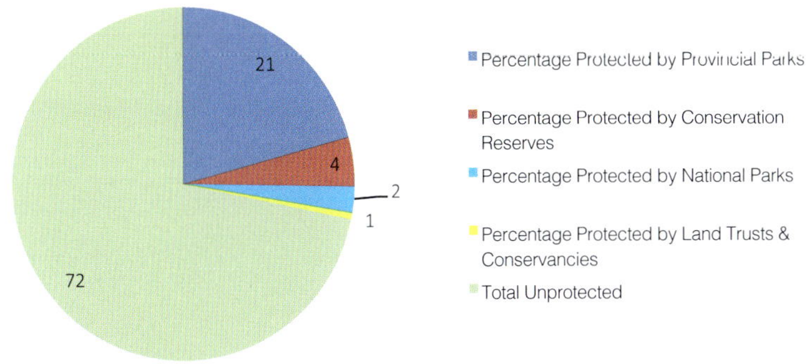

Percent of protected and unprotected spaces in Georgian Bay (2016)

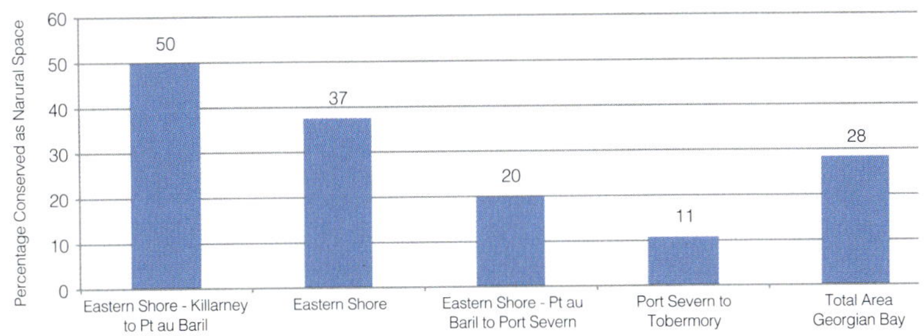

Percent of conserved open and natural places by location in eastern Georgian Bay (2016)

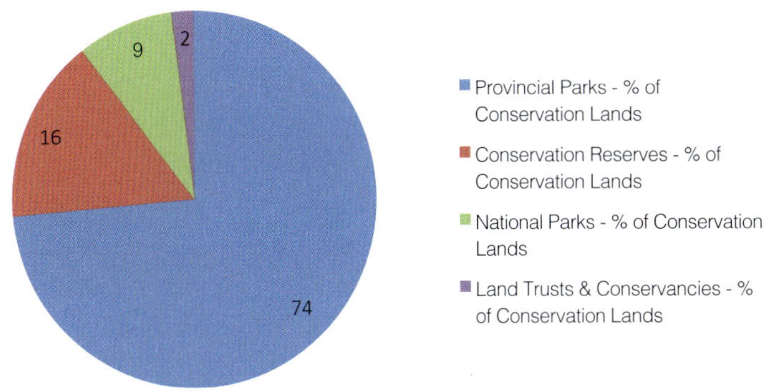

Percent of conserved spaces by conservation mechanism (2016)

The 1980s experienced the largest growth with the creation of almost 100,000 ha of ecologically protected public space. The entire eastern shore of Georgian Bay west of Highway 69/400 (including the Thirty Thousand Islands), and from Port Severn to Killarney, forms a "Natural Area" defined by the Nature Conservancy of Canada and Georgian Bay Land Trust. Later in 2004 much of this area and the entirety of the Bruce Peninsula (1990) were designated UNESCO World Biosphere Reserves.

The 147,000 ha of provincial parks represent some 74% of the total of protected lands (some 21% of the total natural area of Georgian Bay). Conservation Reserves and National Parks currently protect 4% and 2% of the Bay's natural area respectively. The greatest pressures on natural areas are not surprisingly in the south while much of the Thirty Thousand Islands of the Bay's eastern shore is for the most part not protected. While the total area protected by conservancies and land trusts on Georgian Bay is presently only 1% of the total, this will grow considerably should individual landowners continue to make conservation a priority.

National and Provincial Parks of Georgian Bay

In Canada, there are four avenues for preserving landscapes and their habitats in perpetuity. These are: national parks; provincial parks; conservation reserves; and land trusts and conservancies. While Provincial Policies and Municipal Official Plans provide other methods for shorter term protection, these methods see continual pressure and are open to economic and political pressures.

Killarney Provincial Park

The Canada National Parks Act (2000) governs parks and an Act of Parliament is required for national park formation. The mandate is to maintain the natural and cultural heritage features of the park area in an unimpaired state for future generations: "The national parks of Canada are hereby dedicated to the people of Canada for their benefit, education and enjoyment, subject to this Act and the regulations, and the parks shall be maintained and made use of so as to leave them unimpaired for the enjoyment of future generations." Each national park must have a Management Plan "containing a long-term ecological vision for the park, a set of ecological integrity objectives and indicators and provisions for resource protection and restoration, zoning, visitor use, public awareness and performance evaluation." Any development in the national parks must be authorized by Parks Canada, subject to the provisions of the Act.

Georgian Bay Islands National Park was established in 1929 with an area of 13.5 km^2. From the late 1800s through the 1920s the southeastern shoreline of the Bay had seen very rapid settlement. By 1925 Beausoleil Island was one of the last large islands that had not been developed in the southern Thirty Thousand Islands. Cottagers from Honey Harbour and Cognashene saw benefit in preserving open space and petitioned the Canadian government to make Beausoleil Island a national park. And so it was in 1929 that Georgian Bay's first national park was established, though "in early Park days, firewood was cut from within the Park; wolves and porcupines were shot; mosquitoes and weeds were heavily sprayed." Some 63 small island parcels that stretch from Beausoleil north to Twelve Mile Bay have been added since. The park is a refuge for an abundance of flora and fauna and, from an ecological standpoint, is in an important transition zone situated at the southern fringe of the Canadian Shield. Taken as a whole, the park encapsulates many of the landscapes and ecosystems found around the eastern and southern margin of Georgian Bay. It provides significant habitat for colonial waterbirds and migratory bird sanctuaries during spring and fall migrations. Southern Beausoleil Island on the other hand has richer, deeper soils rotted in glacial sediments which support thick maple and oak forests and altogether is very different from the northern end of the island with its bare granite, thin soils, and classic white pine features. The island literally straddles the boundary between the Shield and southern lowlands.

Bruce Peninsula National Park (156 km^2) was only established in 1987 and accounts for 70% of the total area conserved on the Bruce. In the mid-1800s, the "Bruce" was Saugeen Ojibway Nations territory. The plentiful fish stocks and forests drew industry here and the first sawmill was established in 1881. By 1900 most of the big trees and most of the forest were gone. Numerous wild fires sprung up as a result of deforestation, leaving most of the Bruce barren limestone plains by the mid-1920s. It is an outstanding example of a human-made landscape now protected as a "natural area."

Provincial parks are widely distributed around Georgian Bay and range in size from the largest — Killarney (50,000 ha) established in 1964 and which is the only park on the Bay intended to reserve significant areas of wilderness — to small parks such as Sturgeon Bay (14 ha) and Craigleith (66 ha) that are primarily intended for recreation. French River Provincial Park was set up in 1989 to preserve the historic waterway and Shield landscape of an intricate maze of narrow inlets and bare granite bedrock; a canoeist's or kayaker's paradise. One of the most popular parks is Killbear at the mouth of Parry Sound which has campsites right along the shore. Being very accessible from Highway 400, and close to the amenities of Parry Sound, Killbear is traditionally the first introduction many families have to the islands of Georgian Bay and an important jumping off point for kayakers. Other parks are relatively remote, such as the Limestone Islands which protects a limestone plain island and associated bird habitat offshore. Recent work identified one thousand tern nests thought to represent about 2.5% of the total North American population. The total area protected by provincial parks is some 146,404 ha (361,770 acres).

French River

Early summer pollen

Some First Nations have established their own First Nations Parks on Georgian Bay but they are not safeguarded by the same statutes as provincial or federal parks. Cape Croker Park opened in July 1967, a credit to the Chippewas of Nawash Band who recognized the potential and intrinsic value of such an area to Band members. Point Grondine Park is located between Killarney and the French River. Regulated Conservation Reserves are very recent additions to land conservation efforts on Georgian Bay. These areas protect natural and cultural features while providing opportunities for a variety of compatible traditional activities (e.g. fishing, hunting and trapping). They are regulated under the Provincial Parks and Conservation Reserves Act; they also provide crucial places for scientific research and environmental monitoring.

A Conservation Plan for the Future

Another avenue for conserving land around Georgian Bay is through trusts and conservancies such as the Georgian Bay Land Trust, the Nature Conservancy of Canada, the Bruce Trail Conservancy, and the

Escarpment Biosphere Conservancy. These organizations allow private landowners the opportunity to preserve terrestrial spaces and wetlands by donation. Their importance in conserving wild spaces on Georgian Bay now and over future decades cannot be underestimated. The land trusts and conservancies on Georgian Bay work collaboratively with private landowners, township planners, cottage associations, and community groups to effectively conserve Georgian Bay's key biodiversity features and wilderness places. In that regard, the Government of Canada recognizes the extreme importance of natural spaces and accordingly, through its Ecological Gifts Program, provides income tax credits to landowners of ecologically sensitive land who commit to maintaining a portion (or all) of their property in a natural state, by agreement with a land trust or conservancy.

The central aspect here is improved knowledge of the key ecological sites around the Bay. As related above, a key milestone in this regard was the completion of the *Ecological Survey of the Eastern Georgian Bay Coast* in 2001 by Jalava et al. (2005) which was a summary of inventory findings focusing on some 70 natural areas throughout the study area. This was integrated with earlier biophysical studies of the area into an overview of the whole coast. According to its authors, "It reflects the best available scientific information and current principles of conservation biology as they relate to natural ecosystems. Field ecologists will lament the limitations of time and funding, and the challenges in accessing certain data and research, but none of this detracts from the results of this survey, which stand as baseline documentation – and a challenge – to future generations of students of nature. Finally, the report also suggests ways that organizations and individuals can contribute to a sustainable future for this unique landscape."

As a consequence of the work of Jalava et al. (2005), the Natural Area Conservation Plan (NACP) for Eastern Georgian Bay was developed by the Nature Conservancy of Canada with the Georgian Bay Land Trust as a partner. Its objectives are to conserve high priority shoreline, wetland, rock barren and forest ecosystems, and habitat for turtles, reptiles, and migratory fishes on private lands, support research and monitoring initiatives to improve biodiversity and conservation information, and to provide landowner groups and land and water-use planning initiatives with sound biodiversity and conservation information to support stewardship and conservation efforts of wetlands, islands, shorelines of different types, rock barrens, fish, birds, and reptiles.

Future Challenges

Conservation around Georgian Bay has a long and mostly successful history. As always, however, there is much left to be done in the face of relentless urban development along the Bay's southern fringes and increasing numbers of people who are drawn north to a year-round four-season lifestyle. There are, at present, considerably more than 20,000 acres (8,095 ha) of ecologically important lands on the eastern shore of Georgian Bay that are unprotected and which could see development. Some 80% or more of the southern parts of the Bay also remain developable land close to rapidly growing urban communities fuelled by new retirement communities and the desire to live on the water year round. Past practice has largely been to preserve areas of "wilderness" and assume that the integrity and ecological functioning of the Bay will somehow be protected—but this approach is increasingly viewed as too simplistic. First, much of Georgian Bay's landscapes have been overprinted by human activity, especially large-scale logging and second, is it the right type of "wilderness" that is being protected? In this regard, a key concern going forward is whether areas that underpin the broader ecological functioning of the Bay are being preserved in sufficient numbers, and that these areas have the appropriate interconnectivity with each other. Also, is there sufficient understanding of how different aquatic and terrestrial species interact other than anecdotal evidence? What are the main migration patterns of these species and what controls them? What are viable sizes for protected areas needed to ensure their natural ecological functions? What too, represents sustainable land-use practices in Georgian Bay? How resilient are its varied ecosystems to incremental development year upon year as its reputation as a place of solitude becomes known world-wide? These are the key scientific questions.

It is amazing in this day and age that so little is known of the Bay's greatest resource of all — its fresh waters. As related in preceding chapters lake circulation patterns, both seasonal and long term, are a key control on ecological interactions between west and eastern portions of Georgian Bay, but studies on the appropriate scale remain few in number. Is current policy sufficient to ensure the future integrity of the Bay's waters in 100 years' time? Intuition says it's not.

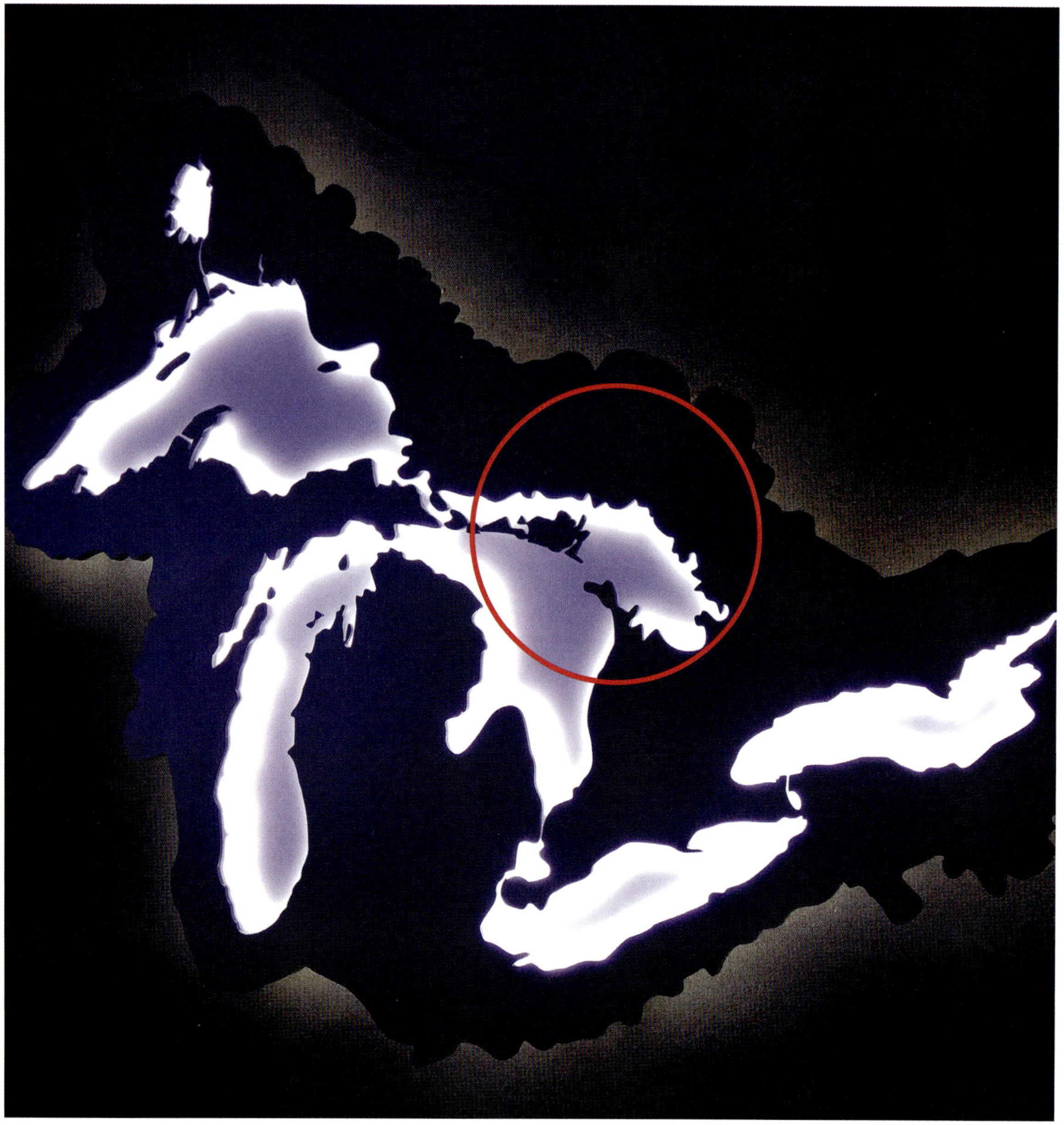

What is an appropriate balance between the concerns and interests of seasonal visitors and residents? Are our public agencies sufficiently equipped to deal with overall planning and the need for scientific data to underpin policy decisions?

A major challenge is to preserve not just the ecologically significant areas of the Bay but in addition those sites that best record its long geological history and the full range of its classic Shield and limestone landscapes. As we have seen, Georgian Bay is a window into the working of our planet and, through the work of Tuzo Wilson and others, can be considered the "birthplace of plate tectonics." Georgian Bay as a landscape has taken more than 600 million years to evolve and has experienced numerous ice ages. The emergence of wind farms, large cottages on increasingly remote islands, and the associated visual blight on classic landscapes around the Bay highlight the need for increased emphasis on landscape protection. A new and growing concern is the need to protect and upkeep (often at great cost) important cultural artifacts such as lighthouses for example, that record key steps in the evolution of human settlement and land use in the area.

To meet these challenges requires much more scientific knowledge about the workings of the Bay than is currently available. It also demands that such knowledge, often hidden away in reports and databases, is transferrable to those who make policy. Given the relatively slow pace at which scientific knowledge is gained and then converted often piecemeal into policy, the rate of development will always exceed that at which our knowledge base increases. The result is that policy tends to be "reactive" rather than proactive. In practice we need to monitor the Bay in real time but this awaits new technologies that can measure the movement of birds and fish from one storm to another, from one season to another. Herein lies the importance of private initiatives through land trusts such as the Georgian Bay Land Trust and others, with their ability to move relatively quickly to secure areas at risk.

Georgian Bay receives many visitors each year, most of whom take advantage of the recreational facilities that its many islands and waterways offer. A key aspect is public education and the generation of a wider awareness regarding the uniqueness of the place we call Georgian Bay and how it came to be over the last 2 billion years before humans entered the scene. We must also recognize that only when appropriate scientific and policy initiatives are set in play can we meet the future challenges of protecting the unique freshwater ecosystems and landscapes of Georgian Bay while respecting the aspirations of indigenous peoples, visitors and residents alike.

The historic Pointe au Baril Lighthouse named after the barrel placed on the point (at left) to guide mariners.

The Great Gneiss Road. The many rock cuts along Highway 400 north of Port Severn provide an unparalleled view of the inner workings of our planet, specifically processes at work deep below mountain ranges.

BIBLIOGRAPHY

Barry, J.P. 1968. The Georgian Bay: the Sixth Great Lake. Clarke Irwin & Co., Ltd., 189pp.

Barry, J.P. 1992. Georgian Bay: An Illustrated History. Boston Mills Press. 233pp.

Bartram, E. 2014. Rockscapes of Georgian Bay. Fitzhenry & Whiteside, Markham, 132pp.

Bastedo, J. Shield Country: the Life and Times of the Oldest Place on the Planet. Red Deer Press, 270pp.

Bennett, E.B. 1988. On the physical limnology of Georgian Bay. Hydrobiologia 163, 21-34.

Bigsby, J.J. 1824. Notes on the Geography and Geology of Lake Huron. Geological Society of London Transactions 1: 204-5, 2: 175-209.

Bogue, M.B., 2001. Fishing the Great Lakes: An Environmental History 1783-1933. University of Wisconsin Press.

Brooks, G. R., Medioli, B. E., 2012. Sub-bottom profiling and coring of sub-basins along the lower French River, Ontario: insights into depositional environments within the North Bay outlet. Paleolimnology 47, 447-467.

Captain Mac, 1884. The Muskoka Lakes and Georgian Bay: Portraying the fishing and hunting grounds, islands and summer resorts, with scenic views and descriptions. Grip Print and Pub, Co.

Comfort, C. 1951. Georgian Bay Legacy. Canadian Art 8.3. Spring (1951), 106, 109.

Culshaw, N. G., D. Corrigan, J. W. F. Ketchum, P. Wallace, et al. 2004. Georgian Bay geological synthesis, Grenville Province; explanatory notes for preliminary maps P.3548 to P.3552.Open File Report - Ontario Geological Survey: 285-28, 5 sheets.

deVisser, J. and Ross, J. 1992. Georgian Bay. The Boston Mills Press. 120pp.

Debicki, R.L. 1982. Geology and Scenery Killarney Provincial Park area, Ontario. Ministry of Northern Development and Mines Geology Guidebook No.6.

Dineva S., Eaton, D.W., Mereu, R. and Ma, S. 2007. The October 2005 Georgian Bay (Canada) earthquake sequence: Mafic dykes and their role in the mechanical heterogeneity of Precambrian crust. Bulletin Seismological Society America 97: 457-473.

Doughty, M., Eyles, N., Wallace, K.W., Boyce, J.I., and Eyles, C. 2014. Lake sediments as natural seismographs: earthquake-related deformations (seismites) in central Canadian (Ontario and Quebec) lakes produced by reactivation of Precambrian structures. Sedimentary Geology 313, 45-67.

Easton. R.M., 1992. The Grenville province and the Proterozoic history of central and southern Ontario. Geological Survey of Ontario Special Vol. 4, Pt. 2, 714-902.

Eyles, N. 2002. Ontario Rocks: 3 Billion Years of Environmental Change. Fitzhenry and Whiteside, Toronto. 352 pp.

Eyles, N. and Miall, A.D. 2007. Canada Rocks: A Geologic Journey. Fitzhenry and Whiteside, Toronto. 568 pp.

Eyles, N. 2011. Canadian Shield: The Rocks that Made Canada. Fitzhenry and Whiteside, Toronto. 128 pp.

Eyles, N. 2012. Rock drumlins and megaflutes of the Niagara Escarpment, Ontario, Canada a hard bed landform assemblage cut below the Saginaw-Huron Ice Stream. Quaternary Science Reviews 55, 34-49.

Eyles, N. 2013. Road Rocks Ontario: More than 250 Geologic Wonders to Discover. Fitzhenry and Whiteside, Toronto. 580 pp.

Eyles, N., Zajc, A. and Doughty, M. 2015. High-resolution seismic sub-bottom reflection record of low Hypsithermal levels in Ontario lakes. Journal of Great Lakes Research 41, 41-52.

Eyles, N. and Doughty, M. 2016. Glacially-streamlined hard and soft beds of the paleo-Ontario Ice Stream in central Canada. Sedimentary Geology 338, 51-71.

Floren, R. and Gutsche, A. 1998. Ghosts of the Bay: A Guide to the History of Georgian Bay. Lynx Images, 302 pp.

Georgian Bay Tourist Company of Midland, 1927. The 30,000 Islands, Georgian Bay, Canada's Summer Playground. Boat Trips between Midland and Parry Sound.

Grady, W. 2011. The Great Lakes. The Natural History of a Changing Region. Greystone Books, Vancouver. 351pp.

Gronewold, A.D., V. Fortin, B. Lofgren, A. Clites, C.A. Stow, and F. Quinn. 2013. Coasts, water levels, and climate change: A Great Lakes perspective. Climatic Change 120: 697–711.

Guidebook and Atlas of Muskoka and Parry Sound Districts. 1879. H.R.Page and Co. Ltd, Toronto.

Hamilton, J.C., 1893. The Georgian Bay: An account of its position, inhabitants, mineral interests, fish, timber and other resources, with map and illustrations: papers read before the Canadian Institute.

Harper, J.R. 1967. Landscape painting in Upper Canada. In: Portraits of a Province. Studies in the History of Ontario. Ontario Historical Society, Toronto, 215-224.

Hietikko-Hautala, T. 2011. The Kvarken Archipelago World heritage Site. Echoes of the Ice Age. Centre for Economic Development for South Ostrobothnia. 158 pp.

Hough, J.L. 1962. Lake Stanley, A low stage of Lake Huron indicated by bottom sediments. Geological Society of America Special Bulletin, 73: 613-620.

Jackson, L.J., Ellis, C., Morgan, A.V., McAndrews, J.H., 2000. Glacial lake levels and eastern Great Lakes palaeo-Indians. Geoarchaeology Int. Journal 15, 415–440.

Jalava, J.V., Cooper, W.L. and J.L. Riley. 2005. Ecological Survey of the Eastern Georgian Bay Coast. Nature Conservancy of Canada, 180pp.

Johnson, M.D., Armstrong, D.K., Sanford, B.V., Telford, P.G. and Rutka, M.A. 1992. Paleozoic and Mesozoic Geology of Ontario. In: Geology of Ontario, Ontario Geological Survey, Special vol. 4, Part II, p. 907-1010.

Ketchum, J.W.F. and Davidson, A. 2000. Crustal architecture and tectonic assembly of the Central Gneiss Belt, southwestern Grenville Province, Canada: a new interpretation. Canadian Journal of Earth Sciences 37, 217-234.

Killan, G. 1992. Ontario's Provincial Parks, 1892-1993: 'We make progress in jumps.' In: Islands of Hope: Ontario's Parks and Wilderness. Edited by L. Labatt and B. Littlejohn. Firefly Books Ltd. Willowdale, Ontario. Pp. 20-44.

Kirkwood A., and Murphy, P. 1878. Undeveloped lands in Northern and Western Ontario: Information Regarding Resources, Products and Suitability for Settlement. Undeveloped lands in Northern and Western Ontario.

Kor, P.S.G. and Cowell, D.W. 1998. Evidence for catastrophic subglacial meltwater sheetflood events on the Bruce Peninsula. Canadian Journal of Earth Science 35, 1180-1202.

Labatt, L. and B. Littlejohn (Editors). 1992. Islands of Hope: Ontario's Parks and Wilderness. Firefly Books Ltd. Willowdale, Ontario. 286 pp.

Lehtinen, M., Nurmi, P.A., Rämö, O.T. (Eds.) 2005.Precambrian geology of Finland: Key to the evolution of the Fennoscandian Shield. Developments in Precambrian Geology 14, Elsevier, 736 pp.

LePan, Douglas. 1998. Towards the Open: Georgian Bay Poems, with wood engravings by Alan Stein. Church Street Press, Parry Sound.

Lewis, C.F.M., Karrow, P.F., Blasco, S.M., McCarthy, F.M.G., King, J.W., Moore Jr., T.C., and Rea, D.K. 2008b. Evolution of lakes in the Huron basin: deglaciation to present. Journal of Aquatic Ecosystem Health & Management 11, 127-136.

Logan, W.E. 1857. On the Division of the Azoic Rocks of Canada into Huronian and Laurentian. The Canadian Journal, July, 439-442.

Luukkola, P., 2003. Journey into Finland. Otava Publishing Co. Ltd. 135pp.

Lytwyn. V.P. 1990. Ojibwa and Ottawa fisheries around Manitoulin Island. Historical and geographical perspectives on aboriginal and treaty fishing rights. Native Studies Review 6, 1-30.

Macfie, J. 2004. The Great North Road. The story of an Ontario Colonization road. Boston Mills Press, 160pp.

McCarthy, D. 1991. The Good Wine: An Artist comes of Age. Macfarlane Walter & Ross, Toronto.

McCarthy, F.M.G., Tiffin, S.H, Sarvis, A.P., McAndrews J.H. Blasco, S.M. 2010. Early Holocene brackish closed basin conditions in Georgian Bay: microfossil evidence. Journal of Paleolimnology 47, 429-445.

McCarthy, F.M.G., McAndrews, J.H., 2010. Early Holocene drought in the Great Lakes basin caused hydrologic closure of Georgian Bay. Journal of Paleolimnology. 47, 411-428.

McGuffin, G., 2003. Great Lakes Journey: Exploring the Heritage Coast. McClelland and Stewart, 144pp.

Meehan. O.M. The Georgian Bay and Great Lakes Survey, 1883-1903.2004. The Northern Mariner XIV No. 1 (January), 7-42.

Murray, F.B. 1967. Agricultural settlement on the Canadian Shield: Ottawa River to Georgian Bay. In: Portraits of a Province. Studies in the History of Ontario. Ontario Historical Society, Toronto, 178-186.

National Oceanic and Atmospheric Administration (NOAA) www.ngdc.noaa.gov/mgg/greatlakes/huron.html.

Pollard, W.C. Life on the Frontier. A sketch of the Parry Sound Colonies that Settled near Edmonton. NWT in the Early Nineties. Arthur Stockwell, London, England. 112pp.

Report of Committee for Ottawa and the Georgian Bay Territory 1864. Legislative Assembly of Canada, printed by Hunter, Rose and Lemieux, https://archive.org/stream/.../reportofcommitt00cana_djvu.txt

Richar, G.A., 1987. Parry Sound 1887-1987 Historical Miniatures. Morsel Press, Nobel, Ontario. 129pp.

Robertson, J.A. and Card. K.D. 1988. Geology and Scenery: North Shore of Lake Huron Region. Ministry of Northern Development and Mines Geological Guidebook No. 4.

Satterly, J. 1942. Mineral Occurrences in Parry Sound District. Ontario Department of Mines Fifty First Annual Report v. LI, Part II, 87pp.

Shaw, M.A. 1929. The Happy Islands: Stories and Sketches of the Georgian Bay. McClelland and Stewart Ltd, 254pp.

Sheng, J. and Rao, Y.R. 2006. Circulation and thermal structure in Lake Huron and Georgian Bay. Continental Shelf Research 26, 1496-1518.

Stone, K. 2008. Paddling and Hiking the Georgian Bay Coast. Boston Mills Press. 256pp.

Storck, P. L. 2004. Journey to the Ice Age: Discovering an Ancient World. University of British Columbia Press. 384pp.

Straw, A. 1968. Late Pleistocene glacial erosion along the Niagara Escarpment of southern Ontario. Geological Society of America Bulletin 79, 889-910.

Trebitz, A. S. 2006. Characterizing seiche and tide-driven daily water level fluctuations affecting coastal ecosystems of the Great Lakes. Journal of Great Lakes Research 32, 102-116.

Tucker, A. 1978. Steam into the Wilderness: Ontario Northland Railway 1902-1962. Fitzhenry and Whiteside, Markham, ON. 215 pp.

Tushingham, A.M. 1992. Postglacial uplift predictions and historical water levels of the Great Lakes. Journal of Great Lakes Research. 18, pp. 440-455.

Verma, H. 1971. Geology and Fossils of the Craigleith Area. Ontario Geological Survey Guidebook No. 7.

West, G.F and others. 2014. J. Tuzo Wilson: A man who moved mountains. Canadian Journal of Earth Sciences 51, xvii-xxxi.

Williams, D.E. 2013. Hardscrabble. The High Cost of Free Land. Dundurn Press.

Wodicka, N., Ketchum, J.W.F. and Jamieson, R.A. 2000. Grenvillian metamorphism of monocyclic rocks, Georgian Bay, Ontario, Canada: implications for convergence history. Canadian Mineralogist 38: 28, 471-510.

Zeller, S. 1987. Inventing Canada: Early Victorian Science and the Idea of a Transcontinental Nation. University of Toronto Press. 356pp.

'In the end, it's all about
those fantastic rocks....'
R. Doktor, 2015

AUTHOR BIOGRAPHIES

Christine Boyanoski is a curator and writer based in Toronto and holds a PhD from the University of London, UK. She has organized many exhibitions and published extensively in the field of Canadian art. Christine is currently writing an online book about the Russian emigrée artist Paraskeva Clark for the Art Canada Institute, and working with the Art Gallery of Hamilton. Another book in progress is a critical examination of the cultural connections between Canada, Britain, and the other former settler colonies in the 1920s and 1930s.

Martin Cooper has been involved in archaeology in Ontario for over thirty years having received an Honours BA in Anthropology from the University of Western Ontario and an MA at the University of Toronto. He is a Partner and Senior Archaeologist with Archaeological Services Inc. and has served as Project Director on hundreds of single and multi-phased assessments throughout Ontario. His special areas of expertise include the archaeology and ethnohistory of northern Ontario and he is an accomplished canoeist and traveller throughout Canada's north.

Nick Eyles is Professor of Geology at the Scarborough campus of the University of Toronto (since 1983) and has published research on the geology of modern and ancient ice sheets from the Arctic to the Antarctic. He has written 5 best-selling books including *Canada Rocks* (with Andrew Miall) and was host of the very popular Canadian Broadcasting Corporation's five-part *Geologic Journey-World* series which aired on the *Nature of Things* in 2010. Nick has been honoured with the Geosciences in the Media Award by the American Association of Petroleum Geologists, the McNeil Medal by the Royal Society of Canada and the E.R. Ward Neale Medal by the Geological Association of Canada. Nick cottages at Carling on Parry Sound in Georgian Bay.

Susan Hannon is a retired ecology professor from the University of Alberta. Her research career focused on behaviour and ecology of birds and the impacts of forestry and agriculture on bird communities. She taught courses in Ecology and Conservation Biology. Currently she lives on Salt Spring Island, BC, where she spends her time volunteering with conservation organizations and trying to restore her property to a Garry oak meadow. She recently spent a winter in Parry Sound helping to care for her mother, and writing about the large mammals of Georgian Bay.

Jamie Hunter has a Master's degree in Museum Studies and has more than fifty years of archaeological and historical experience. He found his first arrowhead on Beausoliel Island in 1964 which triggered a life-long pursuit of studying the heritage of the First Nations, French, British and Canadians in Georgian Bay. Jamie has curated many exhibitions, overseen numerous publications, raised significant funds, and worked for the protection of heritage sites. He is now retired and works as a heritage consultant with his wife Rosemary and lives near Midland, overlooking Georgian Bay.

Bill Lougheed is a retired scientist and biomedical engineer with a life-long interest in conservation on Georgian Bay. Bill's great-grandfather, along with other University of Toronto professors, established the Great Lake's first biological field station on Georgian Bay in 1901. Bill is President of his local Georgian Bay cottage association and is involved in township official plans with a focus on preserving the Bay's wilderness places. He is the Executive Director of the Georgian Bay Land Trust and summers on Georgian Bay.

Andrew Miall has been Professor of Geology at the Department of Earth Sciences, University of Toronto, since 1979, where his focus is teaching and research on the stratigraphy and sedimentology of sedimentary basins. He is the inaugural holder of the Gordon Stollery Chair in Basin Analysis and Petroleum Geology, founded in 2001. He was elected a Fellow of the Royal Society of Canada in 1995 and was President of the Academy of Science from 2007-2009. From 2000-2004 Andrew served as Canada's representative to the NATO Science and the Environment Program's "Committee on the Challenges of Modern Society" and in 2010-2011 served on expert panels struck by the Government of Canada and the Alberta Government to examine the environmental management of the Alberta Oil Sands. Miall has been the author of five research-level technical books and the editor of five special research collections. In 2007 he published *Canada Rocks* co-authored with Nick Eyles.

Karl Schiefer PhD., is a widely recognized specialist in freshwater ecosystems having held academic and research positions focusing on aquatic ecosystem research in Canada and abroad. His expertise includes aquatic ecosystem processes, fisheries ecology and management, limnology and water quality, fish habitat protection, restoration and enhancement, and environmental impact assessment. He has completed projects for both public sector and private sector clients in the fields of damage assessment, Fisheries and Clean Water Act litigation, and environmental planning and impact mitigation. He is an active member of several professional and environmental organizations and advisory committees.

Peter Storck has a Ph.D in anthropology and is Senior Curator Emeritus at the Royal Ontario Museum following a distinguished career on New World archeology and the peopling of North America. He is the author of the award-winning *Journey to the Ice Age: Discovering an Ancient World*, a compelling story of Paleo-Indian peoples in what is now Ontario at the close of the last glaciation. Along with Patrick Julig, and under the overall direction of Ron Williamson, he led the first exploration of the famous Sheguiandah site on Manitoulin Island, a National Historic Site since 1954 and which is almost 10,000 years old. Peter's most recent book, *Fatal Flaw*, is a novel combining archaeology and science fiction and set in the near future.

Janny Vincent is a graduate of the University of Toronto and a second generation islander on Georgian Bay and has spent over 50 years summering with her family there. She shares her island with her husband, daughters, son-in-laws, grandchildren and beloved dogs. She resides in Toronto and is the President of Vincent Associates Inc., a full service information management company and President and C.E.O. of Fundata Canada, one of Canada's leading supplier of financial information. She is Chair of the Georgian Bay Land Trust.

Bob Whittam has a Master's degree and is a well-known ornithologist; he and his wife Judy have many years of experience working on and writing about the birds of Georgian Bay and many other sites across the Great Lakes. He is author of *The Birds of Georgian Bay* and has run a Breeding Bird Survey at Byng Inlet for more than 25 years. He is a former Executive Director of the Friends of Wye Marsh and helped reintroduce the Trumpeter Swan to the area and establish the Severn Sound Remedial Action Plan. He also helped start the Wye Marsh Wildlife Festival in 1986 and has been the resident naturalist on CBC's radio show Ontario Morning.

Ronald F. Williamson is founder and managing partner of Archaeological Services Inc., a cultural resource management consulting firm based in Toronto and Burlington, Ontario. He is an Associate Member of the Graduate Faculty at the University of Toronto (Anthropology) and at Western University (Anthropology) in London and Chair, Board of Directors at the Museum of Ontario Archaeology at London. He holds a PhD in Anthropology from McGill University and has published extensively on the pre-contact and colonial history of the Great Lakes region.